INFORM

Burning Garbage in the US

Practice vs. State of the Art

Marjorie J. Clarke, Maarten de Kadt, Ph.D.,
and David Saphire

Editor: Sibyl R. Golden

INFORM, Inc.
381 Park Avenue South
New York, NY 10016-8806
Tel (212) 689-4040
Fax (212) 447-0689

Printed in the United States of America

Library of Congress Cataloging-in-Publication Data

Clarke, Marjorie J.
Burning garbage in the US : practice vs. state of the art / by Majorie J. Clarke, Maarten de Kadt, David Saphire; editor, Sibyl R. Golden.
p. cm.
Includes bibliographical references and index.
ISBN 0-918780-49-7 : $47.00
1. Incineration--United States. 2. Incinerators--United States--Case studies. I. Kadt, Maarten de. II. Saphire, David. III. Golden, Sibyl R. IV. Title. V. Title : Burning garbage in the United States.
TD796.C57 1991
628.4'457'0973--dc20 91-28771
CIP

INFORM, Inc., founded in 1974, is a nonprofit research organization that identifies and reports on practical actions for the protection and conservation of natural resources and public health. INFORM's research is published in books, abstracts, newsletters and articles. Its work is supported by contributions from individuals and corporations and by grants from over 40 foundations.

Printed on recycled paper

CONTENTS

LIST OF TABLES

PREFACE

The words "burning garbage" all too frequently fan the flames of the solid waste management debates raging in communities across the country, raising the temperature of the discussion without contributing to resolving the issues. Yet the need to find solutions to the very real garbage crisis confronting this country is increasingly urgent. More than 130 communities have already opted for some use of incineration, nearly 100 more have garbage-burning plants in various stages of planning or construction, and dozens of others are debating the value of incineration.

In *Burning Garbage in the US: Practice vs. State of the Art,* INFORM neither advocates nor condemns incineration. Rather, we have aimed at clarifying, for all those concerned about or interested in this strategy, both the environmental impacts of current incinerator operations and the steps that can be taken to maximally reduce the environmental and health threats incineration processes pose.

This study provides the first clear and comprehensive look at the state of the art in garbage burning — the equipment and planning and operating practices that lead to the lowest air pollution, the least toxic ash, and the lowest volume of ash — and the extent to which actual, operating garbage-burning plants measure up to this.

One of our most striking findings is that the state of the art in incineration is not limited to end-of-the-pipe emissions control and ash management, the focal points of federal and state government attention. Rather, it encompasses the whole process from planning and design through worker training, monitoring environmental impacts, and ash handling and disposal.

Of primary significance is the role that planning plays in achieving the best possible environmental performance for incinerators: measuring and categorizing the waste stream; developing an overall community waste management plan that includes source reduction and recycling before designing for incineration; determining which materials remaining in the waste stream should not be burned because they are recyclable, noncombustible, or contain toxic materials or pollutant precursors; sizing the plant appropriately; projecting the amount of ash that will need to be landfilled; and identifying adequate landfill space.

Information is the key to making these important planning decisions. Yet, the research for this report revealed a wide information gap. The lack of easily accessible information about air emissions (amount and type) and ash generation, and the absence of standard measurement formats for emission levels, make definition of the impact of garbage-burning facilities and their comparison to state-of-the-art standards extremely difficult.

Costs are another factor of enormous importance to communities evaluating the choices open to them as they plan their solid waste management strategies. While we have by no means conducted a complete economic analysis, we have identified the key financial factors that must be examined when incineration is considered, and have included information on the construction and operations and maintenance costs incurred by the facilities in this study.

Throughout the United States, people are anxious about the operations of existing waste-to-energy plants in their communities and are debating to what extent incineration should be used as one component of an overall integrated solid waste management strategy. Whether they are individual citizens, members of environmental and community groups, state and municipal officials, government regulators, or members of the waste management industry, we hope this report will help them.

It provides the facts about state-of-the-art practices and technologies that people need if they are to promote the cleanest possible incineration. It also identifies information gaps that need to be addressed if incinerator operations are to be planned and assessed with maximum resource conservation and protection of both the environment and public health in mind. Finally, its comparative analysis and profiles of 15 individual plants provide the groundwork for individuals and communities concerned with evaluating existing garbage-burning facilities.

We hope the discussion of the broad scope of the state of the art, current industry practices, and the fragmented nature of available data will stimulate swift consideration by government, industry, and environmental leaders of the best ways to provide communities and planners with the information they need.

— *Joanna D. Underwood*
President

Acknowledgments

We would not have been able to complete this work without the help and advice of a very large number of people, both inside and outside INFORM. We are grateful to all of them.

Many people shared their time with us and provided us with the information we needed to carry out this study. Plant officials took us through each plant. Workers at the plants responded to our questions. Company officials supplied additional information and reports. Local government officials and representatives from community groups took the time to talk with us. State and local officials responded to our many phone inquiries. In some cases, community group members, in addition to giving us their perspective on local plant operations, shared the comforts of their homes with us.

In addition, we are grateful to the people who supplied us with general technical and regulatory information. Papers by James Donnelly (Manager, Environmental Technologies, Davy McKee Corporation) and Jack Lauber (air toxics engineer at the New York State Department of Environmental Conservation) contributed to the discussion of state-of-the-art incinerator technology. Staff of the Air and Waste Management Division of the United States Environmental Protection Agency (Region II) provided INFORM with position papers interpreting the Agency's air and ash regulations.

We are also extremely appreciative of the time and effort of the reviewers who read and critically commented on drafts of this report. We considered all their comments and incorporated many into the final text. Thanks to: Rhoda Becker (Commissioner of Planning and Research Development, Town of North Hempstead, NY), Nevin Cohen (Manhattan Borough President's Office), Christopher J. Daggett (former Commissioner, New Jersey Department of Environmental Protection), Richard Denison (Senior Scientist, Environmental Defense Fund), Donald A. Drum, Ph.D. (Butler County Community College, PA), Floyd Hasselriis, P.E. (Doucet & Minka, Peekskill, NY), Michael Herz (Cardozo School of Law, New York), Barry Mannis (Vice President, Morgan Stanley & Co.), Parker Mathusa (New York State Energy Research and Development Authority), and David R. Woolley (Executive Director, Center for Environmental Legal Studies).

Within INFORM, many of our colleagues contributed significantly to the research, writing, and production of this report. Joanna D. Underwood, INFORM's President, was in on this project's conception. Her support and encouragement moved it through its several stages to its conclusion. Thanks also to Bette Fishbein, Nancy Lilienthal, and Wally Wentworth for their own critical organizational and editorial skills.

A particular note of thanks is due Sibyl R. Golden, Director of Research and Publications. Her skillful editing and questioning sharpened the study's focus and direction, producing an eminently accessible analysis of a complex technical subject and making this report a tool for *anyone* who wants to understand waste-to-energy plants in the United States today.

This book has been many years in the making, and we are grateful to former INFORM employees Allen Hershkowitz, Ellen Poteet, and Catherine Grant who provided substantial research during early phases of the study.

For turning our massive and complex manuscript into this book, we thank Diana Weyne for her copy editing and proofreading and Elisa Last for her design and electronic publishing skills.

Finally, we thank the Mary Reynolds Babcock Foundation, the Robert Sterling Clark Foundation, Inc., the Geraldine R. Dodge Foundation, The Fund for New Jersey, The New York Community Trust, The Marilyn M. Simpson Charitable Trust, and the Wallace Genetic Foundation, Inc., for their generous financial support of this work.

While we could not have compiled and analyzed the information in this book without the assistance of all these people and institutions, the findings and conclusions are the sole responsibility of INFORM.

CHAPTER 1: INTRODUCTION

Producing energy by burning garbage once seemed to be a savior — first to the energy crisis of the 1970's and then to the municipal solid waste crisis of the 1980's. Over the past two decades, more than 100 waste-to-energy (also called resource recovery) plants have been built in the United States. In 1990, some 84,246 tons of garbage per day were burned in a total of 128 heat-recovering incinerators, up from 25,923 tons per day in 1986 and a mere 990 tons per day in 1970. An additional 19 plants were under construction, and 70 were in the planning stages.[1]

Yet, even as this fast-paced construction has proceeded, as municipalities have sought to meet their solid waste challenges, public and government concerns about the environmental and health impacts of garbage-burning plants have been increasing. These concerns focus on the nature and effects of air emissions, the potential toxicity of ash, and the contribution of burning fuels to global climate change. There is also growing awareness of the environmental and economic benefits of two other solid waste management strategies: source reduction — generating less garbage in the first place — and recycling. These two strategies take on a special importance since the United States carries the unfortunate title of one of the world leaders in per capita garbage production, even as hundreds of landfills that have accepted solid waste close each year. How have we gotten into this situation and what do current trends bode for the future of waste-to-energy plants in the 1990's and beyond?

The United States Garbage Crisis

Four converging trends over the past several decades have contributed to the growing solid waste crisis in the United States. Specifically, the population has increased dramatically, ever more garbage is generated on a per capita basis, waste is growing in toxicity as well as in quantity, and traditional disposal options — landfills — are disappearing.

The first and most basic factor is the growth of the United States population. There are simply many more people to generate waste. In 1950, the United States population was 150 million, less than two-thirds of what it is today. Our 250 million people produce a lot of garbage — some 179,600,000 tons in 1988, or 492,000 tons daily. Piled on a football field and tightly compacted, one year's worth of United States garbage would reach a height of 26 miles.

Second, on a per capita basis, each of these 250 million people is generating more garbage than citizens of most other countries. Our total waste translates into 4.0 pounds per day for each man, woman, and child in the country, almost three-quarters of a ton per person per year. (In some areas, the rate of garbage generation is considerably higher than this national average — 7.3 pounds per person per day in New York City, for example.)

This average rate of garbage production is one of the highest in the world, exceeding that of Japan, West Germany, Sweden, and many other industrialized countries that share our standard of living. And it has been growing: the per capita rate has increased from 2.6 pounds per person just 30 years ago. By the year 2010, per capita garbage generation is projected to increase to 4.9 pounds a day, for a total of 250 million tons per year.[2]

A third factor is what makes up our garbage. Garbage, or municipal solid waste, is defined as the solid portion of the wastes generated by households, institutions, government, and commercial establishments. It includes food and yard waste, all sorts of paper, plastics, glass, metals, wood, rubber, textiles, and some construction waste, but not agricultural, industrial, hazardous, or mineral waste.

More and more of municipal solid waste contains toxic or hazardous constituents — a result of the explosive growth of the United States' synthetic organic chemical industry since the mid-1940's, as well as the increasing use of metals and other pollutant precursors in products. The chemical-based products of this industry have become an integral part of our lifestyle: household cleaning materials, paints, pharmaceuticals, pesticides, and much more. The potential of toxic constituents in such everyday products to enter the environment must be considered when disposal options are evaluated.

Organic chemicals are also used in the production of plastics, an ever-growing portion of our waste stream. In 1988, plastics made up 8 percent of the United States waste stream by weight, or 14.4 million tons, up from 0.5 percent of the waste stream and 0.4 million tons in 1960. Yet, only 1 percent of all plastics were recycled in 1988.[3] And, by volume, plastics now constitute nearly 20 percent of the waste that must be disposed of in the United States.

The fourth and final factor contributing to this country's solid waste crisis is the disappearance of landfill space, which has been the main means of waste disposal. In 1988, 73 percent of United States solid waste was landfilled, compared to 14 percent that was burned and 13 percent recycled.

But landfills are rapidly being closed, either because they are filled to capacity or because of concern over the contamination of groundwater by hazardous substances leaching from the fill sites. New landfills are becoming increasingly difficult to site. Some of the critical statistics are these:

- Between 1978 and 1989, the number of landfills operating in the United States decreased from more than 20,000 to 6600, according to the National Solid Waste Management Association.[4]
- The Environmental Protection Agency projects that more than 2000 more will close in the next five years.

- Since 1980, only 50 to 200 new municipal landfills have been built each year, compared to 300 to 400 a year in the early 1970's.

With landfills closing and garbage production growing, municipalities are increasingly confronted with the problem of how to handle their solid waste.

The Strategy of Integrated Solid Waste Management

Garbage experts (including those in the United States Environmental Protection Agency, New York State, environmental organizations, and the solid waste industry) now agree that an integrated solid waste management plan, based on a hierarchy of strategies, is most desirable from environmental, economic, and health perspectives. Such a plan relies on, first, reducing to the greatest extent possible the amount and toxicity of garbage produced (waste prevention or source reduction); second, recycling or composting everything else feasible; and third, disposal: incineration (only after removing materials that do not burn cleanly or well) or landfilling.

It should be emphasized that integrated waste management does not mean a little bit each of source reduction, recycling, incineration, and landfilling. The language of integrated waste management can be misused to justify waste plans that focus prematurely or predominantly on incineration or landfilling before significant efforts have been made to reduce waste at source and recycle. Indeed, this study underscores the need for planning that emphasizes both source reduction and recycling.

With the exception of source reduction, implementation of this hierarchy has been especially effective in Japan. In 1987, Japan recycled as much as 50 percent of its waste stream, burned about 34 percent, and buried 16 percent.[5] INFORM documented Japan's model garbage management practices (including extensive household separation of wastes, removal of hazardous and noncombustible materials from waste to be incinerated, and sophisticated incinerator emissions control equipment and operational and monitoring practices) in its 1987 report, *Garbage Management in Japan: Leading the Way.*[6]

INFORM's Study: A Focus on Incineration

This INFORM study has focused on the strategy of incineration neither to give it full endorsement nor to condemn it. Rather, recognizing that many municipalities are concerned about the operations of their existing waste-to-energy plants and that others are considering incineration as one component of an overall integrated solid waste management strategy during the planning stages, INFORM undertook this study to identify the state of the art for such facilities — the equipment and practices that produce the best environmental results. A second INFORM goal was to provide perspectives on current incinerator performance by comparing actual operating plants to the state of the art and to each other.

Table 1-1 The 15 Study Plants

State Location	Plant	Official Name
California		
• Los Angeles (City of Commerce)	Commerce	Commerce Refuse to Energy Facility
Delaware		
• Pigeon Point	Pigeon Point	The Delaware Electric Generating Facility
Florida		
• Miami	Dade County	Dade County Resource Recovery Facility
• Lakeland	Lakeland	The McIntosh Power Plant
• Tampa	Tampa	McKay Bay Refuse-to-Energy Facility
Maine		
• Auburn*	Auburn	Auburn Energy Recovery Facility
• Biddeford	Biddeford/Saco	Maine Energy Recovery Facility
Maryland		
• Baltimore	Baltimore	Baltimore Refuse Energy System Company Waste to Energy Plant
Mississippi		
• Moss Point	Pascagoula	Pascagoula Energy Recovery Facility
New Hampshire		
• Claremont	Claremont	New Hampshire/Vermont Solid Waste Project
New York		
• Albany	Albany	Sheridan Avenue Refuse-Derived Fuel Steam Plant
• Fulton	Oswego	Oswego County Energy Recovery Facility
• Peekskill	Westchester	Westchester County Refuse Energy System Co.
Oklahoma		
• Tulsa	Tulsa	Walter B. Hall Resource Recovery Facility
Oregon		
• Brooks	Marion County	Marion County Solid Waste to Energy Facility

* *Closed in February 1990.*

Type	Year Operations Began	Burning Capacity (tons/day)
Mass burn	1987	330
Mass burn and RDF	1987	600
RDF	1982	3000
RDF (90% coal)	1983	500 (RDF)
Mass burn	1985	1000
Mass burn	1981	200
RDF	1987	607
Mass burn	1985	2250
Mass burn	1985	150
Mass burn	1987	200
RDF	1982	600
Mass burn	1986	200
Mass burn	1984	2250
Mass burn	1986	1125
Mass burn	1986	550

The Concept of the State of the Art

The concept of **state of the art,** as used in this book, is defined in two ways. First, it refers to the best technologies and operating practices for reducing the environmental impacts of waste-to-energy incinerators. Second, it refers to the best regularly attainable emissions levels for certain air pollutants. The state of the art has been improving over time as technologies and practices have become more sophisticated, environmental concerns have increased, and regulations have become more stringent. However, the state of the art is a technical, rather than a regulatory judgment. Since technical advances have outstripped regulatory responses, the current state of the art goes well beyond what laws, regulations, and permits in the United States now require.

The goal of state-of-the-art incineration of municipal solid waste is to burn, as cleanly as possible, garbage that cannot be reused, recycled, or composted, thereby producing the least air pollution, the least toxic ash, and the lowest amount of ash.

INFORM's 15 Study Plants

Table 1-1 lists the 15 plants that INFORM studied. Built between 1981 and 1987, they are located in 10 different states and range in capacity (size) from 150 to 3000 tons of waste a day. The plants are of two basic types: **mass burn incinerators** that burn garbage as received with minimal effort on-site to remove objects that may or may not burn well or at all, and **refuse-derived fuel (RDF) incinerators** that burn wastes that have been processed to a uniform size and have had noncombustible materials removed.

Scope of the Study

INFORM's study is intended to provide readers (whether they are state or municipal officials, members of environmental or community groups, concerned citizens, government regulators, or part of the waste management industry itself) with the information they need to assess the environmental performance of both proposed and existing waste-to-energy plants.

This book presents information on waste-to-energy technology and on INFORM's sample plants from several perspectives.

- Chapter 2 summarizes the major findings of the study on the technological state of the art, on the environmental and economic performance of the 15 plants, and on the industry regulatory environment.
- Chapter 3 describes the technology of waste-to-energy plants, discusses the environmental impacts of garbage burning, and examines state-of-the-art equipment, practices, and emissions levels. It is designed to give readers a basic understanding of the current practices in incineration of municipal solid waste that generate the least air pollution, the lowest volume of ash, and the least toxic ash. Chapter 3 also discusses the need for retrofitting existing plants to meet state-of-the-art standards and the technologies available for doing so, important information for citizens of the communities hosting the nation's 128 existing waste-to-energy facilities.

- Chapter 4 analyzes the 15 plants INFORM studied (one has closed since the study began), comparing their environmental performance to each other and to the state of the art discussed in Chapter 3. Detailed tables facilitate these comparisons, permitting conclusions about which technologies and practices are most effective in specific operating plants, which problems are largely solved, and which issues remain to be resolved.
- Chapter 5 looks at the economics of waste-to-energy plants: their construction and operating costs and their revenue sources.
- Chapter 6 provides an overview of the complex patchwork of federal, state, and local regulations affecting garbage-burning plants, including the new federal incinerator regulations (New Source Performance Standards), issued by the Environmental Protection Agency early in 1991. It also discusses key questions facing regulators and identifies problems in obtaining useful and usable data about waste-to-energy plants.
- Appendix A consists of profiles of each of the 15 plants that INFORM visited, with data on basic plant characteristics (such as size and fuel type), waste management in the plant's service area, construction planning, design and operations, monitoring and maintenance, environmental performance, regulations, and costs. These profiles give readers in-depth information about individual plants.

While this book is intended for a nontechnical audience, it has been necessary in places to introduce technical terms. These terms are **boldfaced** and defined when they are first used. They are also included in the Glossary in Appendix D. Other appendices discuss the study's methodology and provide a bibliography for further reading.

Notes

1 All 1990 statistics are from *Waste Age,* November, 1990.

2 US Environmental Protection Agency, Office of Solid Waste, *Characterization of Municipal Solid Waste in the United States: 1990 Update,* June, 1990.

3 *Ibid.*

4 National Solid Waste Management Association, "Landfill Capacity in the Year 2000," 1989.

5 INFORM (Allen Hershkowitz and Eugene Salerni), *Garbage Management in Japan,* New York, 1987.

6 *Ibid.*

CHAPTER 2: FINDINGS AND CONCLUSIONS

INFORM examined incineration of municipal solid waste in the United States in order to answer two basic questions:

1. What is state-of-the-art incineration? What technologies and planning and operating practices lead to the cleanest possible incineration? What are the lowest regularly attainable levels of emissions of air pollutants?
2. To what extent do 15 waste-to-energy plants — selected to illustrate the diversity of technologies and other factors — achieve this state of the art?

The study also looked at the economic and regulatory factors that affect incinerator operations and performance. The methodology of the study, including how INFORM identified state-of-the-art technologies, practices, and emissions levels, and how INFORM obtained data about the 15 incinerators, is discussed in Appendix B.

This chapter summarizes the key findings of the study about state-of-the-art technologies and practices and about the specific performance of the 15 sample plants (listed in Table 2-1). It then describes the conclusions derived from these findings.

INFORM's findings fall into eight areas:

1. an overview of achieving the environmental state of the art
2. planning waste-to-energy plants and determining what materials to burn
3. incinerator design and operations
4. air emissions and monitoring
5. ash management
6. worker training
7. economics
8. regulations

Achieving the Environmental State of the Art: An Overview

INFORM found that three steps are necessary for achieving state-of-the-art incineration – that is, for burning, as cleanly as possible, garbage that cannot be reused, recycled, or composted, and thereby producing the least air pollution, the least toxic ash, and the lowest amount of ash. First, wastes that can be recycled, reused, or composted, those that do not burn well, and those that contain toxic materials or pollutant precursors must be kept out of the incinerator, and the incinerator must be sized appropriately. Second, incineration technologies and practices that minimize the formation of pollutants during combustion must be used. Finally, pollution control and ash management systems that maximally prevent pollutants that are produced from entering the environment are needed.

Table 2-1 The 15 Study Plants

State Location	Plant	Official Name
California		
• Los Angeles (City of Commerce)	Commerce	Commerce Refuse to Energy Facility
Delaware		
• Pigeon Point	Pigeon Point	The Delaware Electric Generating Facility
Florida		
• Miami	Dade County	Dade County Resource Recovery Facility
• Lakeland	Lakeland	The McIntosh Power Plant
• Tampa	Tampa	McKay Bay Refuse-to-Energy Facility
Maine		
• Auburn*	Auburn	Auburn Energy Recovery Facility
• Biddeford	Biddeford/Saco	Maine Energy Recovery Facility
Maryland		
• Baltimore	Baltimore	Baltimore Refuse Energy System Company Waste to Energy Plant
Mississippi		
• Moss Point	Pascagoula	Pascagoula Energy Recovery Facility
New Hampshire		
• Claremont	Claremont	New Hampshire/Vermont Solid Waste Project
New York		
• Albany	Albany	Sheridan Avenue Refuse-Derived Fuel Steam Plant
• Fulton	Oswego	Oswego County Energy Recovery Facility
• Peekskill	Westchester	Westchester County Refuse Energy System Company
Oklahoma		
• Tulsa	Tulsa	Walter B. Hall Resource Recovery Facility
Oregon		
• Brooks	Marion County	Marion County Solid Waste to Energy Facility

* *Closed in February 1990.*

Type	Year Operations Began	Burning Capacity (tons/day)
Mass burn	1987	330
Mass burn and RDF	1987	600
RDF	1982	3000
RDF (90% coal)	1983	500 (RDF)
Mass burn	1985	1000
Mass burn	1981	200
RDF	1987	607
Mass burn	1985	2250
Mass burn	1985	150
Mass burn	1987	200
RDF	1982	600
Mass burn	1986	200
Mass burn	1984	2250
Mass burn	1986	1125
Mass burn	1986	550

None of the 15 plants INFORM studied achieved state-of-the-art performance in all of these areas.

Planning Incinerators/Determining What To Burn

State-of-the-art planning for incineration involves determining which materials to burn and which not to burn, sizing the plant correctly, measuring and categorizing the waste stream, including source reduction and recycling plans in both sizing and financial calculations, projecting the amount of ash that will need to be landfilled, and identifying adequate landfill space.

1. **Materials to be burned.** A state-of-the-art waste management strategy would design for the maximum amount of source reduction and recycling within the community before incineration or landfilling, and for separation of materials that are unsuitable for incineration from the wastes to be burned. These activities preserve natural resources, improve incinerator efficiency, and minimize air emissions, and ash quantity and toxicity.

   A well defined source reduction plan identifies not only an overall percentage reduction goal, but also the baseline amount of waste generated, the time frame for achieving the reduction, and specific strategies for achieving the overall goal. A good starting point would be a cap on per capita waste generation within 5 years, followed by an actual 5 percent reduction over the next 5 years. In developing strategies for reaching such a goal, communities would find it helpful to analyze the various components of their waste stream, so they can focus their efforts both on the largest contributors and on components that could be reduced with minor alterations from current behavior.

   While overall recycling goals for communities will also vary depending on the actual composition of the waste stream, recycling plans for individual materials can reasonably, at a minimum, include 90 percent of yard waste, 85 percent of beverage containers (glass, metal, and plastic), 85 percent of newsprint, corrugated, and office paper, and 20 percent of all other paper (goals set by the state of New Jersey). Materials that are unsuitable for burning include noncombustible wastes (such as bottles and cans), those that are explosive (such as gasoline or fuel for camping stoves), and those containing toxic substances or pollutant precursors (such as batteries, some plastics, and yard waste). In many of these cases, materials may be simultaneously suitable for recycling and unsuitable for incineration. A variety of technical studies have demonstrated considerable reduction in pollutant production when presorting of wastes occurs.[1]

2. **The necessity of correct sizing.** Since combustion is most efficient when an incinerator consistently burns the quantity and quality of garbage it was constructed to burn, a general, overriding principle for designing a solid waste incinerator is to determine the correct size. Sizing is based on the amount of heat the boiler can handle. Different components of municipal solid waste have different heat values, frequently measured in British Thermal Units (BTU) per

pound; for example, approximately 13,000 BTU/pound for plastics, 6000–7000 BTU/pound for paper and wood, none for metals and glass. Thus, the composition of the waste is as important to define during planning as its volume.

3. **Measuring and categorizing the waste stream.** To identify the appropriate size for a plant, it is also necessary to carry out waste composition studies that determine the amount and composition of the waste stream, including identifying toxic materials or pollutant precursors and materials that can be targeted for source reduction or removed for recycling. The optimal time for such studies is immediately prior to the plant design process, with samples from different neighborhoods at different times of the year. It is also important to consider future population growth, employment trends, and waste generation, as well as plans for increased recycling and source reduction in the future.

4. **Financial implications of sizing decisions.** Since the financial arrangements for an incinerator may include long-term (15-30 years) contracts for the quantity of garbage delivered to the plant or the quantity of energy to be sold, it is vital that communities plan for source reduction and recycling before an incinerator is built. Otherwise, such long-term contracts may preclude a community's undertaking aggressive source reduction and recycling later on.

Looking at INFORM's 15 sample plants, neither recycling nor source reduction was considered during the planning process for any of these incinerators, although recycling now exists in most of the communities served by the plants in this study. Only four of the plants (Biddeford/Saco, Commerce, Pascagoula, and Pigeon Point) conducted elemental analysis of the waste stream that could reveal the presence of toxic materials or pollutant precursors prior to incinerator construction.

Incinerator Design and Operation

State-of-the-art incinerators are designed to facilitate identification and removal of unacceptable wastes and to maximize combustion efficiency.

1. **Identifying and removing prohibited wastes.** Ideally, unacceptable wastes would be separated, and recycled if possible, before reaching the incinerator, through source separation programs in homes, businesses, and institutions. Screening for unacceptable wastes that reach the burn plant is best accomplished by designing the screening area so that waste can be closely inspected and removed if necessary. Tipping floors, which resemble large warehouse floors, are better suited for this than pits. At refuse-derived fuel plants, mechanical processing and sorting are included in the preparation of the incinerator's fuel.

2. **Designing for greatest combustion efficiency.** Uneven conditions in the furnace, particularly temperature variations, contribute to inefficient combustion and greater production of pollutants, as do insufficient turbulence of combustion gases and insufficient time for burning them. Thus, design features and operational practices that maintain temperatures at optimal and stable levels,

retain combustion gases in the furnace for appropriate times, and create sufficient gas turbulence are desirable. These include proper mixing and drying of heterogeneous wastes prior to burning to make their energy values and moisture content more homogeneous; continuous rather than batch loading to permit a more even flow of fuel to the furnace; avoiding frequent start-ups and shut-downs; and using auxiliary burners, air distribution systems, and automatic combustion controls.

Although all 15 plants in INFORM's study reported methods for some screening of wastes brought to the facility and for identifying prohibited wastes, the amounts of waste actually rejected are small. Of the 12 mass-burn plants, only four reported rejecting 1 percent or more of the incoming wastes (Auburn, Commerce, Oswego, and Pigeon Point) and none rejected more than 3 percent. Considering that 15.5 percent of the United States waste stream by weight consists of metals and glass,[2] much more waste could be captured for recycling or rejected simply on the criterion of noncombustibility. All the mass-burn plants that rejected more than 1 percent of incoming wastes have tipping floors for screening, rather than pits.

Penalties for bringing prohibited wastes to the plants are varied, and the extent of enforcement is limited. Only two plants, for example, Commerce and Dade County, have the authority to levy fines on violators who bring prohibited wastes to the facility; neither has ever levied such a fine. Three plants (Auburn, Lakeland, and Westchester) have no penalties at all.

INFORM observed wide variation in the combinations of types of furnaces, loading techniques, pollution control systems, and auxiliary equipment at the 15 plants in the study (described in Chapter 3 and itemized in the plant profiles in Appendix A). These many different structural features, operational practices, and human actions interact in complex ways that make it difficult to draw correlations between individual factors and reported emissions levels. While theory and experimental tests may suggest that certain designs are superior to others, INFORM's study technique did not permit the identification of causal relationships.

Air Emissions

For the purpose of this study, INFORM has identified state-of-the-art emissions levels for six key air pollutants, defined state-of-the-art monitoring practices, explored pollution control equipment and operating practices that reduce emissions, and compared the performance of the 15 sample incinerators to these state-of-the-art standards.

State-of-the-Art Emissions Levels

INFORM has identified state-of-the-art emissions levels for six key incinerator air pollutants: particulates, dioxins and furans, carbon monoxide, sulfur dioxide, hydrogen chloride, and oxides of nitrogen. The levels were chosen based on a broad review of plant permit limits, recent test reports from operating incinerators, technical and professional reports, regulations, and recommendations of environmental regulatory agencies, as detailed in Chapter 3. INFORM's primary criterion was that the emissions level selected for each pollutant has been achieved in practice, with regularity, using currently available technology. It is important to note that, in many cases, these currently attainable state-of-the-art levels are considerably more stringent than emissions limits mandated by existing federal and state regulations.

Pollutant	**State-of-the-Art Emission Level**
Particulates	0.010 grains per dry standard cubic foot
Dioxins and furans	0.10 nanograms per dry normal cubic meter (Eadon toxic equivalents)
Carbon monoxide	50 parts per million
Sulfur dioxide	30 parts per million
Hydrogen chloride	25 parts per million
Oxides of nitrogen	100 parts per million

INFORM also reported on emissions of two heavy metals (lead and mercury) but did not identify state-of-the-art levels for them because there was not enough information available on regularly attained low emissions levels of these materials.

Monitoring and Measurements

State-of-the-art practice includes two methods for monitoring air emissions and plant operations: continuous monitoring of both emissions and operational factors and periodic measurement of stack emissions for pollutants that cannot be reliably or meaningfully measured on a continuous basis.

Continuous monitoring

State-of-the-art continuous process monitors (CPMs) and continuous emissions monitors (CEMs) track the performance of an incinerator at all times so that, in the event of combustion upsets or high emissions of one or more pollutants, corrective measures can be implemented in a timely manner. Thus, these monitors differ from stack testing that assesses emissions levels at specific times. As identified by INFORM, state-of-the-art continuous monitors measure nine operating and emissions factors: furnace and flue gas temperature, steam pressure and flow, oxygen, carbon monoxide, sulfur dioxide, oxides of nitrogen, and opacity (a crude measure for particulates). Continuous monitoring of hydrogen chloride emissions is also possible, and may soon be (although it is not yet) a sufficiently widely accepted technique to be considered a state-of-the-art requirement.

Telemetering, or instantaneous computer transmission of continuous monitoring data to local or state authorities, can be an excellent method of ensuring adequate, sustained environmental performance, provided that the results are in fact monitored by the appropriate authorities in environmental protection and public works departments.

None of INFORM's 15 study plants had continuous monitors for all of the seven operating and emissions factors recommended and surveyed by INFORM (because data on flue gas temperature and steam flow were not consistently obtained, these two factors were not included in INFORM's analysis of the 15 plants). Three plants (Commerce, Pigeon Point, and Westchester) had six each. All the plants reported monitoring furnace temperature, and all but one (Biddeford/Saco) reported monitoring steam pressure and oxygen. Only three plants had continuous monitors for oxides of nitrogen (Commerce, Pigeon Point, and Westchester) or for sulfur dioxide (Commerce, Lakeland, and Pigeon Point). Only two plants (Oswego and Biddeford/Saco) telemetered the readings outside the plant.

Stack measurements

Stack measurement patterns at INFORM's 15 study plants were inconsistent and in most cases incomplete. Additionally, the plants did not use standard measurement techniques or units of measurement, thereby hampering comparison of the plants.

1. Only three facilities (Biddeford/Saco, Commerce, and Marion County) measured all six pollutants for which INFORM determined state-of-the-art emissions levels and reported their results in a format that INFORM could compare with other results.
2. Two plants also tested for all six but used measurement units for dioxin/furan emissions (Pigeon Point) or carbon monoxide emissions (Tulsa) that did not permit comparison with the data obtained for the other incinerators.
3. Ten plants measured from one to five pollutants, only some in formats permitting comparison.
4. Nine plants measured heavy metal (lead and mercury) emissions in a way that allowed INFORM to compare them.

Pollution Control Equipment

Add-on emissions control devices are used to neutralize, condense, or collect the pollutants generated in the waste-burning process to prevent them from being emitted into the air, although incinerator design, operating practices, and fuel cleaning (source reduction and separation systems) can significantly reduce the amount of pollutants actually produced in waste-to-energy plants.

1. State-of-the-art control of emissions of acid gases such as hydrogen chloride and sulfur dioxide can be achieved using scrubbers.
2. Scrubbers are crucial, although not alone sufficient, for lowering dioxin/furan emissions; certain operating conditions (especially low flue gas temperature) are also needed.
3. State-of-the-art particulate emissions levels are commonly attainable with fabric filters or four-field electrostatic precipitators.
4. Reducing carbon monoxide emissions is best accomplished through plant practices that promote efficient combustion (such as continuous fuel loading and use of automatic combustion controls and auxiliary burners).

5. State-of-the-art control of emissions of oxides of nitrogen requires both minimizing their formation and using chemical injection control devices to neutralize those that are formed. Flue gas recirculation systems can also be helpful.

Emissions at INFORM's Study Plants

As a group, the 15 study plants fell far short of meeting the state-of-the-art emissions levels identified by INFORM. Because of inconsistent monitoring activities at the plants, systematic across-the-board comparison of the emissions from all 15 incinerators was impossible (see Table 2-2). It is also important to remember that stack tests document emissions at one discrete point in time and do not provide a picture of a plant's overall performance over time; Biddeford/Saco, for instance, achieved state-of-the-art levels for four pollutants but has experienced numerous operating problems, including fires and blowouts of its fabric filter system that spewed ash over the local area.

1. **Only one plant providing comparable data (Commerce) attained INFORM's state-of-the-art emissions levels for all six primary pollutants.**
2. One plant (Biddeford/Saco) achieved state-of-the-art levels for four pollutants (it provided comparable data on all six key pollutants).
3. One plant (Marion County) achieved state-of-the-art levels for three pollutants (it provided comparable data for all six pollutants).
4. Five plants (Baltimore, Claremont, Oswego, Pigeon Point, and Tulsa) attained state-of-the-art levels for only one or two of the six pollutants (they each provided comparable data for four or five pollutants).
5. **Six plants did not achieve state-of-the-art levels for any of the six pollutants:** Albany, Auburn, Dade County, Pascagoula, Tampa, and Westchester (they provided comparable data on from one to five pollutants). (Neither did Lakeland, but state-of-the-art levels identified for garbage-burning plants may not strictly apply to this plant since 90 percent of its fuel is coal.)

The 15 study plants exhibited a dramatic range in emissions levels for the six pollutants for which INFORM identified state-of-the-art levels.

1. **Particulates and carbon monoxide are the best controlled pollutants.**

   Particulates. Seven of the 15 study plants (Baltimore, Biddeford/Saco, Claremont, Commerce, Marion County, Pigeon Point, and Tulsa) reported emissions lower than INFORM's state-of-the-art level of 0.010 grains per dry standard cubic foot. With the exception of the now-closed Auburn plant, all the others had levels ranging up to 2.5 times the state-of-the-art level. The plants that achieved the state-of-the-art levels all had either fabric filters or three- or four-field electrostatic precipitators.

   Carbon monoxide. All but one of the seven plants reporting carbon monoxide emissions achieved INFORM's state-of-the-art level of 50 parts per million

Table 2-2 Air Pollutant Measurement/Emissions Levels

Plant	Particulates	Carbon Monoxide	Dioxins/Furans	Hydrogen Chloride	Sulfur Dioxide	Oxides of Nitrogen
	▲ Pollutants Measured / ■ State-of-the-Art Emissions Level Achieved					
California						
• Commerce	■	■	■	■	■	■
Delaware						
• Pigeon Point	■	■	NC	▲	▲	▲
Florida						
• Dade County	▲	NA	NA	NA	NA	NA
• Lakeland	▲	NA	NA	NA	▲	▲
• Tampa	▲	NC	NA	NA	▲	NC
Maine						
• Auburn†	▲	NA	NA	NA	NA	NA
• Biddeford/Saco	■	▲	■	■	■	▲
Maryland						
• Baltimore	■	■	NA	NA	▲	▲
Mississippi						
• Pascagoula	▲	NC	NA	NC	NC	NC
New Hampshire						
• Claremont	■	■	NC	▲	▲	▲
New York						
• Albany	▲	NA	▲	▲	▲	▲
• Oswego	▲	■	▲	▲	▲	▲
• Westchester	▲	NA	▲	▲	▲	▲
Oklahoma						
• Tulsa	■	NC	▲	▲	▲	▲
Oregon						
• Marion County	■	■	▲	■	▲	▲

NA, Information not provided by plant.
NC, Measurement not comparable with others.
* *Since 90% of Lakeland's fuel is coal, rather than municipal solid waste or refuse-derived fuel, the state-of-the-art levels identified for garbage-burning plants may not apply.*
† *Closed in 1990.*

State of the Art for How Many Pollutants?	Number of Pollutants Measured (in comparable format)
6	6
2	5
0	1
*	3
0	2
0	1
4	6
2	4
0	1
2	5
0	5
1	5
0	5
1	5
3	6

(Pigeon Point, Commerce, Marion County, Oswego, Baltimore, and Claremont); the remaining plant (Biddeford/Saco) reported emissions less than twice the state-of-the-art level. Except for Pigeon Point and Oswego, all of these plants use continuous loading, all have auxiliary burners, and all but Biddeford/Saco have automatic combustion controls.

2. **For the other four key pollutants, no more than three plants achieved state-of-the-art emissions levels for any one of them.**

   Dioxins and furans. Only two of the seven plants that reported dioxin/furan emissions levels in a comparable format achieved INFORM's state-of-the-art level of 0.10 nanograms per dry normal cubic meter, Eadon toxic equivalents (Commerce and Biddeford/Saco). Emissions as high as 188 times the state-of-the-art level were reported (Albany). Both of the plants reporting state-of-the-art dioxin/furan emissions had scrubbers and fabric filters, as did the plant (Marion County) with the next lowest level of emissions.

   Hydrogen chloride. Three of the nine plants reporting hydrogen chloride emissions in a comparable format attained INFORM's state-of-the-art level of 25 parts per million (Biddeford/Saco, Commerce, and Marion County). All of these, and the plant with the next lowest emissions (Claremont), have scrubbers. The plants without scrubbers reported hydrogen chloride emissions ranging from 16 to 25 times the state-of-the-art level.

   Sulfur dioxide. Of the 12 plants reporting sulfur dioxide emissions levels in a comparable format, the three with the lowest emissions, including two that achieved INFORM's state-of-the-art level of 30 parts per million (Biddeford/Saco and Commerce), have scrubbers. As with hydrogen chloride emissions, emissions of sulfur dioxide exhibited a wide range of variation, with the highest reported emissions (389 parts per million at Oswego) 13 times the state-of-the-art level.

   Oxides of nitrogen. Only one of the 11 plants reporting emissions of oxides of nitrogen in a comparable format achieved the state-of-the-art emissions level of 100 parts per million (Commerce); it is also the only plant with an emissions control device specifically designed to chemically reduce oxides of nitrogen. The plant with the next lowest emissions level (Pigeon Point, 115 parts per million) has a flue gas recirculation system (and a dual-chambered furnace) that helps reduce formation of oxides of nitrogen.

3. **Emissions of lead and mercury, the two heavy metals investigated by INFORM, also varied dramatically.** The highest measured lead emissions level (Albany, 1.28 milligrams per normal cubic meter) is more than 300 times the lowest (Commerce, 0.0042 milligrams per normal cubic meter). The highest reported mercury emissions level (Westchester, 1.92 milligrams per normal cubic meter) is 50 times the lowest (Biddeford/Saco, 0.0448 milligrams per normal cubic meter).

4. **Factors related to achieving state-of-the-art emissions levels included emissions control equipment and plant age.**

   Effectiveness of emissions control equipment. Overall, scrubbers and fabric filters were more effective in reducing air emissions than other emissions control devices: three of the four plants with this combination of equipment are the plants that achieved INFORM's state-of-the-art emissions levels for three or more pollutants (Biddeford/Saco, Commerce, and Marion County). While some plants with electrostatic precipitators attained state-of-the-art levels for particulate emissions, these devices did not reduce emissions of such other pollutants as acid gases and dioxins and furans. Technologies specifically designed for control of emissions of oxides of nitrogen are necessary to achieve state-of-the-art levels. The one plant achieving the state-of-the-art level for oxides of nitrogen emissions (Commerce) was the only one using a chemical process to reduce such emissions; the plant with the next lowest emissions (Pigeon Point) used flue gas recirculation.

   Impact of age of plants. In general, the newer plants reported lower emissions than plants built only a few years earlier: they usually have more technologically sophisticated, and/or more, equipment.

Retrofitting Existing Plants

Retrofitting can enable existing waste-to-energy incinerators with emissions falling short of state-of-the-art levels to reduce their air emissions to varying degrees.

1. **The need for retrofitting.** Since most older United States resource recovery plants were permitted when there were fewer and less stringent emissions control regulations, typical emissions control and combustion equipment for such facilities falls far short of current state-of-the-art technologies. In order to meet increasingly stringent emissions control requirements, including new Environmental Protection Agency guidelines for existing plants, some facilities will be faced with a choice between adding new and/or upgrading existing equipment or shutting down.

2. **Retrofitting technologies.** A variety of post-combustion emissions control technologies have been successfully retrofitted to existing solid waste incinerators, in the United States, Canada, and Europe, allowing them to meet most emissions requirements for acid gases, particulates, dioxins/furans, mercury, and oxides of nitrogen. Technical design considerations for evaluating which retrofit system is most useful for an individual plant include levels of emissions control desired or required, space availability, and system compatibility.

3. **Retrofitting goals.** The cleanest possible incineration would be obtained by requiring existing facilities to retrofit to state-of-the-art standards. These standards exceed both the new EPA requirements for new waste-to-energy plants and the new EPA guidelines for existing plants. Given both the potential

public health impact of incinerator emissions, and the cost of a major retrofit, the question of what should trigger a retrofitting requirement—that is, how far short of state-of-the-art standards a plant must be—is a policy issue deserving serious discussion.

Ash and Its Management

The first priority in state-of-the-art ash management is to reduce both the volume and the toxicity of the residue left after burning of municipal solid waste by removing noncombustibles and materials containing toxic substances from the waste stream before incineration and by ensuring efficient combustion. In Japan, the goal for ash amounts (for incinerators burning more than 200 tons of garbage per day) is 5 percent of the original waste volume. No comparable goal has been established in the United States.

None of the 15 plants INFORM studied verifiably meet this Japanese ash volume goal. Although 12 plants estimated their ash volumes, these figures cannot be considered reliable because incinerators measure the weight of the waste entering the plant and the weight of the ash leaving it, but generally do not measure volumes. Further the volume information provided by many of the plants seemed inconsistent with the weight information they reported.

Turning to ash weight, for which the Japanese have no goal, ash weights ranged from 10 to 50 percent of the weight of the original waste at the 14 plants reporting these data in a way that permitted comparison. Nine of them reported ash weights falling between 20 and 29 percent of total waste. The plant with the lowest ash weight percent, Biddeford/Saco, burns refuse-derived fuel from which, presumably, many noncombustibles have been removed.

State-of-the-Art Ash Management Procedures

State-of-the-art ash management practices are designed to minimize worker and citizen exposure to potentially toxic substances in ash during handling, transportation, treatment, and disposal, long-term storage, or reuse. Such practices are increasingly specified because of the uncertainty that exists in testing ash. While testing is important, there is disagreement about which, if any, of the existing procedures for testing ash toxicity provide reliable information about the potential for toxic substances leaching from ash under realistic landfill conditions. Additionally, sampling and analyzing large volumes of ash on a sustained basis is expensive, and there is no agreement about how to take smaller and less frequent, but still representative, samples.

Safest ash management has several components.

1. The **bottom ash** (noncombustible or partly burned solid material left in the furnace) and **fly ash** (material captured by emissions control devices) is kept separate to allow for more rigorous handling and treatment of the potentially more toxic fly ash.

2. The ash is contained while it is still in the plant.
3. The ash is transported wet in leakproof, covered trucks to disposal sites.
4. The ash is treated to minimize its potential toxic impact.
5. The ash is disposed of in ash-only monofills because codisposal of ash with municipal solid waste may increase the leachability of the ash by exposing it to acid. State-of-the-art monofills have liner systems consisting of multiple layers of composite liners (usually plastic and clay or compacted soil) sandwiched between leachate collection and leak detection systems, with on-site leachate treatment facilities. Ash reuse, as an alternative to disposal and storage, has been a controversial matter with debate continuing over the issues of ash toxicity and human exposure.

Less than half of INFORM's sample plants test their ash regularly, and none uses all of the state-of-the-art ash management practices (handling, transportation, treatment, and disposal) identified by INFORM.

1. **Ash testing.** Only six of the plants in the study provided information on regularly scheduled ash testing, with the remainder reporting occasional testing or testing only during the initial start-up period. Twelve of the plants claimed to use or have used the EPA's Extraction Procedure Toxicity Test (EP Tox) for the toxic content of ash. However, only some states require ash that fails EP Tox tests to be disposed of in a hazardous waste landfill. (In early 1991, the EPA substituted a different ash test, the Toxic Characteristic Leaching Procedure, or TCLP, for the EP Tox test.)
2. **Ash handling and transportation.** None of the 15 plants separated fly and bottom ash. Only four plants (Biddeford/Saco, Claremont, Marion County, and Tulsa) reported covering ash inside the plant *and* transporting it in covered, leakproof containers or trucks.
3. **Ash disposal.** Only two of the 15 plants in the study (Claremont and Marion County) used all of the state-of-the-art disposal techniques identified by INFORM: ash monofill, multiple liners, leachate collection, and leachate treatment (the study did not examine whether they also have leak detection). Two others (Albany and Tulsa) had virtually no protective measures, using neither liners nor leachate collection and treatment. The remaining 11 used some combination of containment techniques.

Future Ash Disposal Capacity

When an incinerator is built, planners need to ensure that adequate disposal space for the ash residue will be available for the expected operating lifetime of 20-40 years.

Five of the study plants were already close to running out of landfill capacity. Specifically, Biddeford/Saco projected reaching its landfill's capacity in 1991 and Marion County in 1995. Albany, Oswego, and Tulsa provided data during INFORM's original research showing they expected to run out of space before 1991; they did not include updated information on this point when responding to INFORM's prepublication follow-up questionnaire. All but the two plants with on-site disposal (Lakeland and Dade County) will reach capacity by 2010, while the incinerators themselves may have additional years of useful life, assuming 30-year operating lifetimes.

Worker Training and Safety

The training of operators is vital to the cleanest possible functioning and best environmental performance of waste-to-energy plants. After a plant's design is optimized, its performance is largely dependent on the quality of operations and maintenance. In a state-of-the-art plant, workers need to understand the effects of variation in the composition and moisture of the waste stream on the combustion system, the emissions control systems, and environmental performance. They must know how to correct upsets and restore steady-state conditions.

Since well-trained specialists are required for optimal operation of complex modern incineration plants, state-of-the-art training for upper-level plant workers (chief facility operator, shift supervisor, and control room operator) involves both formal academic and practical education, as well as supervised on-the-job training. At a minimum, certification would require a four-year bachelor's degree in a technical field, six months of specialized practical training in simulation and other types of laboratory situations, and successful work experience in a resource recovery plant with on-the-job training and close supervision for at least six months in the position. Lower-level employees, such as crane operators, tip floor personnel, and pollution control equipment technicians and mechanics, also need formal and on-the-job training before being certified for specific jobs. Germany, Switzerland, and Japan all have comprehensive training programs for incinerator operators.

Even when the United States Environmental Protection Agency proposed national standards for waste-to-energy incinerators in 1989, they did not include national programs for training of incinerator operators at any level. Thus, these new EPA regulations, which became effective early in 1991, fall far short of a state-of-the-art training program. While they include worker certification standards for chief facility operators and shift supervisors promulgated by the American Society of Mechanical Engineers, they mandate no formal education or training beyond a high-school education. Further, they apply only to the two highest-level plant workers, and they limit recertification to an annual review by employees of a plant-specific operations and maintenance manual.

INFORM found worker training at the 15 study plants to be almost entirely on-the-job. Relatively few workers, including chief facility operators, had any previous garbage burning experience (in part due to the relatively low number of incinerators at which they could have gained such experience).

The Economics of Waste-to-Energy Plants

Incineration economics are complex and variable. While all waste-to-energy plants operate within the same basic framework, each is unique in its individual financial arrangements. They can vary in the public or private nature of their construction, financing, and operation.

The tremendous variability of the costs of ash management and financing, and the reluctance of some plant operators to provide financial data to INFORM, inhibited a detailed economic analysis of the plants in this study. However, data about construction and operations costs and about garbage tipping fees and other revenues do provide a picture of the magnitude and variability of these economic factors.

Construction costs for the 15 plants in the study at the time of construction (excluding financing and not adjusted for inflation) ranged from just under $4 million (Auburn, which started operations in 1981 and has since been closed) to $239 million (Westchester, which started to run in 1984). The cost per ton of design capacity ranges from $19,900 (Auburn) to $110,038 (Biddeford/Saco). To take into account actual plant operations, rather than design capacity, INFORM estimated the total amount of garbage each plant will burn over a 30-year lifetime by multiplying the average amount of garbage each plant is currently burning by 365 days and 30 years. On this basis, construction costs per ton of garbage to be burned over the 30-year expected lifetime of a plant range from $3.64 (Albany) to $12.13 (Westchester).

1. In general, construction costs per ton of design capacity have been increasing over time as a result both of the use of more (and more sophisticated) equipment and of inflation in the cost of labor and materials.

2. Plants that are not currently operating near their design capacity level also show comparatively high costs per ton of garbage to be burned over their lifetime.

Operations and maintenance costs for the study plants (excluding ash management costs) range from $13.33 per ton (Tulsa) to $41.51 per ton (Commerce). While, in general, operations costs do not exhibit any patterns of relationships with age or size of plant, the high costs at Commerce could be due to its extensive emissions control equipment. High per ton costs at Dade County ($35.45) and Pigeon Point ($28.09) can be attributed partly to their currently operating at low percentages of design capacity.

Comparing the environmental performance of the study plants with their costs is possible only in a qualitative way because both the emissions data and the financial information available to INFORM were incomplete. In general, however, the costs of equipping a plant with the most up-to-date pollution control devices contribute to overall construction costs. The only plant to achieve INFORM's state-of-the-art emissions levels for all six pollutants studied (Commerce) is also one of three plants with construction costs of more than $100,000 per ton of design capacity.

With capital costs in the millions of dollars, owners and operators of waste-to-energy plants, as well as providers of construction financing, use a variety of strategies to assure themselves of adequate flows of revenues.

1. **Source of revenues.** Waste-to-energy plants obtain operating revenues from a combination of tipping fees charged for bringing garbage to the plant, sales of electricity or steam, sales of materials (primarily ferrous metals) separated out of the waste stream, and, in some cases, taxes.
2. **Control over revenue.** Plant owners/operators maintain control over revenues through long-term contracts for energy and other products and through flow-control ordinances that guarantee the plant will receive specified amounts of garbage for processing. Financial arrangements developed at the time of planning an incinerator can also include a variety of tax incentives to reduce the amount of debt service and/or taxes the plant operating company must pay.

The Regulatory Environment

Until recently, regulation of the operation of waste-to-energy plants involved a complex patchwork of federal and state standards and individual permit conditions that were occasionally at odds, frequently confusing, and constantly changing. Before the United States Environmental Protection Agency issued incinerator regulations in early 1991, the main federal standards applicable to solid waste incinerators were general ambient air concentration limits for pollutants, with no specific regulations for the emissions from incinerators (except for particulate emissions). In the absence of detailed national incinerator emission standards, some states required plants to meet specific, but constantly evolving, emissions limits. Both sets of regulations were used, case-by-case, to set permit conditions for individual new plants.

The EPA's Standards of Performance for New Stationary Sources (Municipal Waste Combustors), often referred to as the New Source Performance Standards (or NSPS), issued in early 1991, are the first comprehensive national incinerator regulations. They establish overall air emission standards for six air pollutants emitted by incinerators (particulates, carbon monoxide, hydrogen chloride, sulfur dioxide, dioxins and furans, and oxides of nitrogen), define guidelines for good combustion practices, identify monitoring requirements, and specify certification standards. The original proposal also mandated materials separation programs, but the EPA deleted this requirement in the final version. The standards apply to new garbage burning plants with individual combustion units that have the capacity to burn 250 tons per day or more. Existing incinerators with combustion units capable of burning 250 tons per day or more are covered by different, less stringent guidelines, and smaller incinerators will be the subject of future regulations. As a result of the 1990 Amendments to the Clean Air Act, the EPA is required to revise many of the specific provisions of these new regulations by November 1991.

In many respects, the new federal regulations fall short of state-of-the-art standards as identified by INFORM.

1. **Good combustion practices/carbon monoxide emissions.** The new standards set carbon monoxide emissions levels of 50 to 150 parts per million (depending on the type of furnace) as evidence of efficient combustion. The upper limit of this range is considerably higher than INFORM's 50 parts per million state-of-the-art level, a level regularly attained or even exceeded by new state-of-the-art facilities.

2. **Other air emissions.** The regulations establish maximum emission levels or removal efficiencies for particulates, dioxins/furans, hydrogen chloride, sulfur dioxide, and oxides of nitrogen. Most of the proposed limits are less stringent than the state-of-the-art levels identified by INFORM. Further, no specific emissions standards are mandated for heavy metals (although the 1990 Amendments to the Clean Air Act require the EPA to develop such limits within a year), and dioxin/furan standards are stated in a way that does not address the varying toxicity of individual dioxin/furan compounds.

3. **Monitoring.** The regulations establish continuous monitoring for oxides of nitrogen, sulfur dioxide, carbon monoxide, and opacity; and annual stack tests for particulate matter, dioxins/furans, and hydrogen chloride. The state-of-the-art continuous monitors identified by INFORM also include furnace and flue gas temperature, steam pressure and flow, and oxygen; state-of-the-art monitoring would also include stack tests for heavy metals.

4. **Certification.** The regulations specify that chief facility operators and shift supervisors be certified according to standards promulgated by the American Society of Mechanical Engineers. These standards fall short of the state of the art because they include no formal training requirements beyond high school and apply only to the two highest levels of operating staff.

5. **Materials separation.** The originally proposed regulations included a 25 percent materials separation requirement (reduction in solid waste entering the incinerator) that represented a new direction for the EPA, an indication that it intended to actively support its solid waste management hierarchy. However, this requirement was completely deleted from the set of regulations that was finally adopted.

Ash is virtually unregulated on the federal level and regulations are only beginning to be proposed at the state level. The diversity of existing regulations and practices reflects unresolved controversies over ash management, particularly over the classification of ash from garbage-burning plants and the reliability of methods for assessing toxicity.

1. **Assessing toxicity.** There is no agreement about which, if any, of the existing testing procedures provides reliable information about ash toxicity under actual landfill conditions. Without agreement about testing procedures, ash classifica-

tion is difficult, and it is impossible to determine the safety of various ash reuse options.

2. **Ash classification and disposal.** The debate over whether to classify ash as legally hazardous or nonhazardous, or as some intermediate special waste, is fueled significantly by cost. The Resource Recovery and Conservation Act (RCRA) mandates that materials legally classified as hazardous be disposed of in special landfills, following specific regulations, with costs significantly higher than those for disposal in ordinary landfills. However, RCRA has been read to exempt household waste and its products, such as ash, from consideration as legally defined hazardous waste. The issue of establishing procedures for classifying and managing incinerator ash is likely to be debated when Congress takes up reauthorization of RCRA, during its 1991-1992 session.

The regulatory disarray INFORM encountered during the course of this 15-plant study highlights the confusion facing regulators and citizens alike. Key factors include inaccessibility of information and lack of standardization. The new EPA regulations resolve many of these issues, but leave others outstanding.

1. **Inaccessibility of information.** The inaccessibility of information about air emissions and air regulations is one of the most significant factors limiting serious discussion of air pollution issues and resource recovery standards. Without clear, easily available information, presented in a standard format, it is difficult, and in some cases impossible, to compare requirements and environmental performance at individual incinerators to state-of-the-art standards. Specifically, information is inaccessible because it is scattered among a variety of agencies and may be located in different agencies in different states, because regulations are sometimes undergoing revision and review and are typically expressed in specialized terminology that can vary from state to state, and because emissions levels are determined and expressed using a variety of standards, methods, and measurement units.

2. **Lack of standardization.** The overlapping and inconsistent nature of much of the existing regulation has caused permit conditions to vary from state to state and plant to plant, with different pollutants regulated, different pollution management techniques used, and different acceptable emissions levels set. Beyond this, however, there has also been no agreement about the way to express emissions levels: what measurement units to use, what set of operating conditions to define as a standard to which all measurements could be corrected, and what averaging times to use.

3. **Impact of new regulations.** The new national standards resolve many of these issues. By establishing stack emissions limits for six pollutants of concern, they define measurement criteria to be used. Assuming data are made publicly available, this will make it easier to compare the performance of an individual plant to national standards and to other facilities. Some regulatory differences are inevitable, however, since states have the authority to impose regulations stricter than federal standards.

Conclusions

The 15 waste-to-energy incinerators INFORM studied are not operating as cleanly as possible. With few exceptions, they fail to meet state-of-the-art standards for emissions of air pollutants, ash management, monitoring practices, and worker training. Further, the variations in emissions levels are, in many cases, dramatic: for example, emissions of dioxins and furans at the Albany plant were 188 times the state-of-the-art level, and emissions of lead from the Albany plant were more than 300 times greater than lead emissions from Commerce. Although, by definition, the state-of-the-art emissions levels are regularly attainable in practice, existing laws, regulations, and permits do not require incinerators to meet them.

At the same time, public concerns about incineration are growing. Factors contributing to this concern include the nature and effect of air emissions, the toxicity of ash, odors from plants, the diminishing supply of ash disposal sites, and the role burning plays in global climate change. They also include the ever-growing volumes of municipal solid waste created by increasing per capita garbage generation and an increasing population.

In light of these concerns and the findings of this study, INFORM draws several conclusions.

1. To preserve natural resources and to minimize the size of an incinerator, and therefore the amounts of emissions and ash it produces, it is essential to plan incinerator construction only in the context of an overall community waste management strategy that first maximizes source reduction and recycling.

2. To minimize air emissions and the quantity and toxic content of ash, a community waste management strategy needs to also include separation of nonrecyclable, noncombustible materials and materials containing toxic substances or pollutant precursors from the wastes to be burned.

3. Accurate preconstruction size planning includes not only the measurement of the types and quantities of wastes produced in a community, but also an assessment of the impact of existing or future source reduction and recycling on the wastes that will actually be burned. Accurate plant sizing is vital both because an incinerator operates most cleanly and efficiently when it consistently burns the quantity (tons per day) and quality (energy value) of wastes for which it was designed, and because plant owners and providers of construction financing often require guarantees that a plant will receive specified amounts of garbage in order to ensure projected revenue flows. Such preconstruction planning thus reduces the potential for conflicting demands of source reduction, recycling, and incineration programs.

4. The cleanest incineration could well be ensured by regulations requiring incinerators to meet the technically feasible and regularly attainable state-of-the-art pollutant emissions levels; to burn only combustible wastes that do not contain pollutant precursors; and to adopt state-of-the-art ash management,

monitoring, and formal worker training practices. Existing regulations — including the EPA's recently issued Standards of Performance for New Stationary Sources (Municipal Waste Combustors) — do none of these.

5. Mechanisms for encouraging the continuing evolution and improvement of the state of the art need to be developed. The EPA could monitor and disseminate information on the best new techniques. Requiring plants to keep up with the evolving state of the art would provide an economic incentive for companies to develop and use new and better equipment and practices.

6. To best protect the environment and public health, retrofitting the many older waste-to-energy incinerators that lack up-to-date emissions control, combustion equipment, and continuous monitors to meet current state-of-the-art emissions levels would be essential. Yet, the EPA's new regulations will only require existing plants with individual combustion units capable of burning 250 tons per day to meet emissions guidelines, and these guidelines are generally less stringent than the regulations for new plants which, as noted above, fall short of the state of the art.

7. As more sophisticated air pollution control devices capture more potentially toxic pollutants (which otherwise would escape as air emissions) in the fly ash, ash management is becoming increasingly important and ash management regulations are needed. Reliable methods for assessing the toxicity of ash under realistic disposal conditions need to be developed so appropriate transportation, handling, and disposal or reuse methods can be specified.

8. To make possible public comparison of individual plants to state-of-the-art standards, information about the environmental performance of incinerators needs to be standardized and made more accessible. Specifically, air emissions other than those covered by the EPA's new regulations could be reported in a standard format: measurement units, operating conditions that measurements would be corrected to, and averaging times. While more work needs to be done to specify which chemicals to monitor, certain categories can be identified, such as heavy metals and volatile organic compounds. This information could well be gathered and made public annually through one central location in each state and through a national information clearinghouse.

   A model for such a national database is the EPA's Toxics Release Inventory, mandated under the 1986 Superfund Amendments. It provides the public with annual figures on the releases and transfers of some 320 individual toxic chemicals and chemical groups from more than 22,000 industrial waste-generating facilities nationwide.

Notes

1 These studies, including ones conducted for the United States Department of Energy and the Environmental Protection Agency, are discussed in the section "Pollution Prevention Through Waste Presorting" in Chapter 3; sources are identified in footnotes 11-15 for that chapter.

2 US Environmental Protection Agency, Office of Solid Waste, *Characterization of Municipal Solid Waste in the United States: 1990 Update*, 1990.

Chapter 3: The Technology of Garbage Burning

Waste-to-energy incinerators burn municipal solid waste, reducing its volume, recovering energy, producing air emissions, and creating ash in the process. There the similarity among such incinerators ends, for a wide variation exists in the techniques used. This chapter is designed to give readers a basic understanding of the current practices in incineration of municipal solid waste that have the potential to generate the least air pollution, the least toxic ash, and the lowest amount of ash. The operation and performance of the 15 plants surveyed by INFORM are compared to these best, or state-of-the-art, practices later in this study. Governments and community groups concerned about the cleanest waste-to-energy operations in their areas can compare the operations of existing plants, or the plans for proposed ones, to the best practices discussed here.

The key to understanding these best practices is understanding how garbage-burning plants work and how their operations may affect the environment. Thus, this chapter first provides an overview of these topics before the detailed discussion of the state of the art in reducing the environmental impacts of waste-to-energy incinerators.

How Garbage-Burning Plants Work: An Overview

Incinerators reduce the volume of municipal solid waste (garbage) by burning it in an enclosed environment, creating ash and air emissions. Large-scale incinerators may handle as much as 4000 tons of garbage per day, collected from homes, institutions, and businesses. The incinerators discussed in this study are **waste-to-energy** (also called **resource recovery**) plants that are designed to recover energy, in the form of steam, that can be circulated for heating, or converted to electricity. Figure 3-1 shows the outside of a typical garbage-burning plant, the Baltimore facility in this study.

Types of Plants

Two main types of waste-to-energy incinerators have been examined in this study: mass burn incinerators and refuse-derived fuel incinerators. Since many factors that interrelate with each other in complex ways are involved in the operations of waste-to-energy facilities, the type of incinerator — mass burn or refuse-derived fuel — does not by itself determine the plant's overall environmental performance.

The more common **mass burn incinerators** burn garbage as received with minimal effort on-site to separate objects that may not burn well or burn at all. (For example, bulky, oversized items such as tires, bedframes, fences, and logs are often separated by hand to avoid problems, but glass bottles, metal cans, and batteries usually are not.)

Refuse-derived fuel incinerators, on the other hand, burn wastes that have been preprocessed and sorted (either on the site of the incinerator or at a separate processing

Figure 3-1

Exterior of a garbage-burning plant: the Baltimore Refuse Energy Systems Company (BRESCO) Waste to Energy Plant. (Photo: David Saphire)

facility). Recyclable materials such as ferrous metals, aluminum, and glass are separated mechanically and collected for processing and future sale or disposal. The remaining material, called **refuse-derived fuel,** or **RDF,** is then used as a fuel and burned in the incinerator.

Incinerator Structure and Processes

Wide variation, as mentioned above, exists in the practices used for incineration. The basic process, however, is generally constant: garbage is brought to the plant, where it may be separated or preprocessed; then it is burned in a furnace; the heat generated is recovered as energy; and air emissions and ash residues resulting from combustion are managed using various technologies. Figure 3-2 diagrams the basic structure of a typical mass burn waste-to-energy plant.

Moving garbage from delivery truck to furnace

In a typical mass burn incinerator operation, garbage is transported by truck to the incinerator plant. The truck is weighed to determine the actual tonnage of garbage being brought in. The garbage is then discharged either into a pit, as shown in Figure 3-2, or onto a concrete tipping floor, for storage and initial mixing. This is the stage at which any effort to remove bulky, noncombustible, or hazardous wastes, or wastes otherwise unsuitable for incineration, usually takes place.

Wastes are also prepared for incineration at this stage. This preparation can range from basic mixing of wastes in most mass burn plants to intensive separation of materials, shredding, and pulverizing in refuse-derived fuel plants.

From the storage pit or tipping floor, the wastes in a mass burn plant are typically

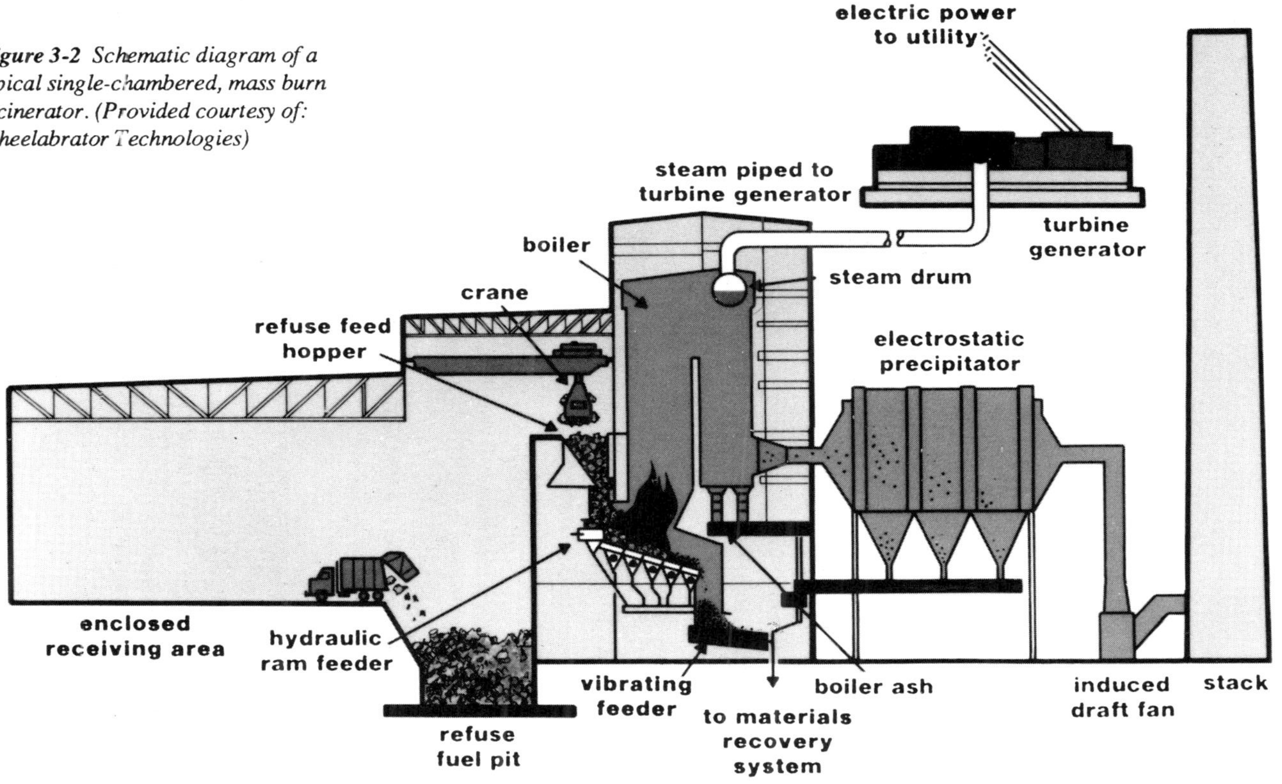

Figure 3-2 *Schematic diagram of a typical single-chambered, mass burn incinerator. (Provided courtesy of: Wheelabrator Technologies)*

transferred by a crane or front-end loader (bulldozer) to a charging chute. The waste moves through the chute onto the first of a series of grates in the furnace that move the garbage through the furnace. In a refuse-derived fuel plant, the fuel is injected into the combustion chamber.

Burning garbage

In the furnace, the heat dries the garbage and subsequently ignites it. Then the garbage burns and, finally, some of the gases formed as the solid waste burns are themselves burned. The burning of garbage optimally involves a reaction in which wastes are broken down, in the presence of heat and oxygen, into gaseous components, thereby releasing energy. This process is called **combustion.**

Complete combustion would reduce the garbage to its simplest form of carbon dioxide, water, and inert residue. However, the extent to which this occurs depends on a variety of factors including the amount of available oxygen, the temperature, the extent to which the garbage and gases are mixed, and the amount of noncombustible materials in the waste. Often garbage is not completely burned or broken down, and intermediate gaseous and particulate products are formed in the incinerator. If they are not trapped and condensed by emissions control devices, these products may be emitted into the air in the form of gases and small particles, or they may accumulate at the bottom of the incinerator as unburned or partially burned solid matter known as **bottom ash.**

Recovering energy

Hot gases generated as a result of combustion exit the furnace and pass through the boiler which recovers the energy in the form of steam. The steam may be sold directly or converted into electricity in a turbine.

Managing emissions and ash

Pollutants, formed at various stages of the incineration process, fall into two categories, air emissions and ash (to be discussed in more detail in the next section, "Environmental Impacts of Garbage Burning"). After leaving the boiler, gases may pass through a series of emission control devices before exiting the plant through a stack; Figure 3-2 depicts one such emissions control device, known as an **electrostatic precipitator**.

Ash includes both matter (char, metals, glass) that is not completely burned on the furnace grate (bottom ash), and matter that leaves the furnace suspended in combustion gases and falls out of the gases as they cool (such as **boiler ash**), or is subsequently trapped in emission control devices (**fly ash**). Figure 3-2 shows bottom ash from the furnace going to a materials recovery system.

Environmental Impacts of Garbage Burning

Though modern incinerators are designed with increasing attention to reducing pollution, and technology has improved in recent years, there are, nonetheless, environmental impacts. Unwanted air emissions include particulate matter, heavy metals, acid gases, oxides of nitrogen, and products of incomplete combustion, including chlorinated organic compounds (e.g., dioxins, furans, chlorobenzenes, and chlorophenols), many of which can be toxic if they enter the body in sufficient concentration. Incinerators, like all combustion devices, also emit considerable quantities of carbon dioxide which, although not toxic, is considered one of the major contributors to the global climate change phenomenon. Additionally, ash is produced; it may have toxic components. Plant operations may also have other environmental impacts, such as the effects of truck traffic on neighboring communities, odors, and discharge of wastewater.

Further, overall energy and natural resource use increases when materials that could be recycled or reduced at the source are burned. Both source reduction and recycling decrease the amounts of energy and resources needed to produce materials.

Air Emissions

Since the garbage entering the furnace is never completely burned or broken down, intermediate products form in the incinerator and may be emitted into the air in the form of gases or small particles, together called **emissions**. Some materials, such as metals, do not break down at all. Many of these noncombustible or partially burned products (either constituents of the original waste item or newly created compounds) can be toxic in certain concentrations. Additionally, some are subject to federal regulations because they have been classified as criteria air pollutants by the Clean Air Act: carbon monoxide, sulfur dioxide, oxides of nitrogen, lead, particulates, and ozone (not directly emitted).

Particulates and heavy metals

The incineration of municipal solid waste produces minute particles in solid or condensible form called **particulates.** Particulates range in size from more than 500 microns to less than 0.1 micron in diameter. (One **micron** equals one one-thousandth of a millimeter or 1/25,000 of an inch. The dot over the letter "i" in this text has a diameter of about 400 microns; it contains approximately 160,000 particles with a diameter of 1 micron.) People can inhale particles smaller than 10 microns. Those less than 2 microns can penetrate the body's natural defense mechanisms and lodge in the deepest, most sensitive parts of the lungs, allowing greater opportunity for uptake into the body.

Municipal solid waste contains numerous metals that cannot be destroyed by burning. So-called **heavy metals,** including arsenic, cadmium, chromium, lead, mercury, and nickel, are released from garbage incinerators during burning. If not controlled, heavy metals may leave the stack on fine particulate matter or as vapor and be discharged into the air; some portion collects in ash. A Canadian study tracked 27 heavy metals from waste to municipal solid waste incinerator bottom ash, fly ash, and emissions.[1] Many of these metals can be toxic and can pose health hazards to humans. For example, chronic low-level lead exposures have been linked to learning disabilities and behavioral problems;[2] mercury is known to damage the central nervous system; and cadmium exposure has been linked to cancer, pulmonary emphysema, and kidney and cardiovascular damage.[3]

Products of incomplete combustion

A variety of compounds jointly known as **products of incomplete combustion** are produced when garbage does not completely burn. Some of these, for example, some of the 75 **dioxin** and 135 **furan** compounds, are known to be highly toxic to certain animals, although very little direct information is known about their effects on humans, or about their impacts under conditions of long or concentrated exposure typical of incinerator emissions.

Carbon monoxide (CO), a common product of incomplete combustion, is a colorless, odorless gas that interferes with the blood's ability to absorb oxygen. Exposure to carbon monoxide in large doses (not typical of most incinerator emissions) can also slow reflexes, impair perception and thinking, cause drowsiness, and, in extreme cases, cause unconsciousness and death. The effects of long-term human exposure to low levels of carbon monoxide, alone or in combination with other pollutants, is unknown. Carbon monoxide is also a significant contributor to the formation of urban smog.

Acid gases

Acid gases form during combustion when certain elements in garbage come in contact with oxygen or hydrogen. **Sulfur dioxide** (SO_2), for example, forms when sulfur and oxygen combine. **Hydrogen chloride** (HCl) and **hydrogen fluoride** (HF) are formed from hydrogen and, respectively, chlorine and fluorine in the refuse. When released into the atmosphere, these gases contribute to acidification of rain or fog, and consequently to metal corrosion (for example, of water supply pipes) and to erosion of limestone and marble buildings and statues.

Oxides of nitrogen (NO, NO_2, collectively termed $\mathbf{NO_x}$) are formed during garbage incineration by the combination of nitrogen from the wastes and the air with oxygen from the atmosphere. There is an urgent need for minimizing emissions of oxides of nitrogen in major cities, southern California, and the eastern megalopolis since oxides of nitrogen reduce visibility, are one of the major precursors to ozone, urban smog, and acid rain, and contribute to respiratory and eye irritations.

Ash

Depending on the design and operation of the plant, the incineration of municipal solid waste can leave substantial and varying ash residues. (One of the key factors in reducing ash amounts is presorting to remove noncombustible materials.) While Japan has established a goal for ash amounts of 5 percent of the original waste volume, volume is barely reported in the United States. However, solid waste and ash are generally weighed, and ash amounting up to 50 percent by weight of the original wastes was reported by the plants INFORM studied.

Types of ash

Ash falls into two main categories — bottom ash and fly ash. The bottom ash is composed of noncombustible and incompletely burned solid material that falls through or is left on the grates as wastes move through the furnace. Its composition varies depending on the content of the waste and the efficiency of combustion.

Fly ash is composed of solid particulates and condensed acids, organics, and metals captured by the emission control devices. The quantity of fly ash collected depends both on the amount of pollutants produced, and on the efficiency of the pollution control devices in condensing and capturing them. The more efficient they are, the more fly ash will be collected.

Ash toxicity

Incinerator ash may be toxic to humans, if they are exposed to it, due to the physical and chemical changes caused by combustion. These changes concentrate pollutants (such as heavy metals) in a smaller volume of waste and permit human exposure by breathing or ingestion.

Knowledge of the chemical composition of ash is central to any assessment of its potential impact on air, surface water, or groundwater. Air impacts result from dispersion of dry ash in handling, transportation, or disposal. As with garbage itself, ash effects on groundwater and surface water supplies are caused by materials leaching from landfills without adequate leachate collection and treatment systems or liners, and by run-off from landfills that lack sufficient stormwater controls.

While metals in the original waste are self-contained in bulky matter (for example, batteries, auto parts, electronics), metals in the ash are collected on tiny particles subject to dispersion by wind or leaching by water. These tiny particles can be easily inhaled or ingested directly by humans. They can also be dissolved in water and, if ash migrates to groundwater and surface waters, absorbed into the food chain.

The question of whether these ash particles pose a threat to human health is controversial. An Environmental Defense Fund study found that the levels of certain metals (among them, lead and cadmium) in incinerator ash, especially fly ash, are high enough to cause severe health effects if people are exposed to them.[4] However, some industry experts believe that the metals cannot leach from the ash, and thus that human levels of exposure are not high enough to present problems.[5]

Ironically, the more efficient the air emissions controls, the greater the problems for ash management. The toxic potential of fly ash is increasing due to the advances in air pollution control technologies that capture greater quantities of emissions and leave the residue in a more concentrated form. The challenge is how to keep the toxic components in ash out of the air, land, and water.

Given the wide variety of incinerator designs and operations, and the variability of the garbage itself, the nature and quantity of toxic constituents in municipal solid waste incinerator ash varies. It is thus not possible to state categorically how potentially toxic ash is. It is even a challenging task to assess the toxicity of ash from a single incinerator, based on an individual ash sample, due to the variability of the waste stream, the lack of widely accepted toxicity tests, and the inconsistency of analytical results.

Other Environmental Impacts

In addition to producing air emissions and ash, waste-to-energy plants can have operational impacts on the environment and neighboring communities. For example, their water use and wastewater handling can affect community water supplies, and truck traffic can create noise and air pollution.

Water usage and disposal

A reliable source of pure water which is used to produce steam is important to the long-term operation of any industrial boiler, including those found in the energy recovery sections of waste-to-energy plants. Even though boiler water is recycled, some additional make-up water is needed to replace losses. Impure make-up water can cause buildup of deposits of impurities on metal duct work. Waste-to-energy plants also use water to cool combustion gases before they pass through pollution control devices, to wash the plant, and to quench the hot ash remaining after garbage is burned. Finally, a large amount of water is evaporated when wet cooling towers are used to condense steam back to water that is fed to the boilers. (Some plants use air, rather than water, in their cooling towers.)

Water for all these purposes is usually taken from municipal water supplies and from nearby bodies of water (rivers, harbors). Where water supplies are insufficient, wet cooling towers and wet scrubbers may not be practical.

Municipal solid waste incinerators in the United States generally do not use wet scrubbers that generate wastewater in large quantities since this wastewater and other effluents must be treated in on-site wastewater treatment facilities or in off-site municipal sewage treatment plants. Some modern plants, such as the Claremont incinerator in INFORM's study, are designed to have "zero discharge" of wastewater.

Traffic

Truck traffic to and from garbage-burning plants is typically of concern to residents living near these facilities because of the potential for ash dispersal from uncovered or leaking trucks, emissions from truck exhaust, and impacts from additional traffic. (Other methods of waste disposal also require truck movements — to and from processing and disposal facilities — but do not present the ash transportation problems.) Some larger incinerators may have a few hundred trucks a day either unloading garbage or removing ash. Some plants will only accept waste at certain times of the day, causing long lines of idling trucks to form.

In order to minimize impacts of truck traffic, it is desirable to design the loading and unloading area within the plant to facilitate easy and rapid unloading. Truck access to the plants should be sufficiently staggered to prevent traffic jams and road widths and traffic control systems (such as traffic lights and signs) should be designed to accommodate traffic levels and minimize accidents.

The State of the Art in Reducing Environmental Impacts

The goal of state-of-the-art incineration of municipal solid waste is to burn garbage as cleanly as possible, producing the least air pollution, the least toxic ash, and the lowest amount of ash. There are three fundamental steps to achieving this: first, keeping wastes that can be reused or recycled, that do not burn well, or that contain toxic materials or pollutant precursors out of the incinerator, and sizing the incinerator appropriately; second, employing incineration technologies and practices that minimize the formation of pollutants during combustion; and, third, using pollution control systems that maximally prevent pollutants that are produced from entering the environment.

Each of the many techniques discussed in the remainder of this chapter affects the degree to which the impacts of garbage burning can be minimized in one of these three ways. Knowledge of what works best, and why, is critical to understanding which existing plants perform well and whether proposals for new plants incorporate state-of-the-art equipment and techniques.

The Concept of State of the Art

The term **state of the art,** as used in this book, is defined in two ways. First, it refers to the best technologies and operating practices for reducing the environmental impacts of waste-to-energy plants. Second, it refers to the best regularly attainable emissions levels for certain air pollutants. Determining the state of the art is a technical rather than a regulatory judgment, although the state of the art could be used as the basis for regulatory standards. Since technical advances have outstripped regulatory responses, the current state of the art goes well beyond what today's United States laws, regulations, and permits require. Whether or not a plant achieves the state of the art can be assessed by the types of equipment, operations, and pollution control technologies it uses, as well as the solid waste materials it burns. It can also be assessed by the air emission levels the incinerator achieves.

The state of the art in solid waste incineration has been improving over time. Although incineration of garbage with energy recovery has been going on, to a limited extent, for several decades, the technologies and practices used in processing the waste stream, in incinerating the waste, and in controlling/managing the emissions and ash output have changed considerably. Over the last decade, as landfill space has become scarce, interest in incineration has been renewed: environmental concerns have increased and regulations have become more stringent. Thus, it is not surprising that, in INFORM's evaluation of 15 plants, emissions from the more modern garbage-burning plants were generally lower than those achieved by plants constructed in 1981.

The sources we have used to define state-of-the-art technologies and practices refer to experience in actual incinerator operations, and the emissions levels we consider state of the art have, in all cases, actually been attained in practice. (Table 3-4, in the section on "Emissions Control," lists some of the plants where state-of-the-art emissions levels have been achieved.)

Designing the Plant

Building a state-of-the-art garbage-burning plant begins with the design. The size of the plant is a critical factor; thus, planners need accurate information about the amount and type of wastes the plant is to burn, as well as a sense of future garbage management practices in the community.

Figure 3-3 Bulky materials removed from waste delivered to the RDF processing plant in Baltimore (not in INFORM's study). (Photo: David Saphire)

Determining what to burn

A state-of-the-art waste management strategy would design for the maximum amount of **source reduction** (reducing the amount and/or toxicity of garbage generated) and **recycling,** including **composting,** within a community before incineration, in keeping with the widely accepted hierarchy of garbage management strategies. Additionally, materials that are not recyclable, but that are unsuitable for burning because they are noncombustible, explosive, or contain toxic substances or pollutant precursors, should be separated from wastes to be burned. These activities preserve natural resources, improve incinerator efficiency, and minimize air emissions and ash quantity and toxicity. Figure 3-3 shows some bulky materials, which probably would not burn well, separated at the Baltimore RDF processing facility.

In order to achieve effective source reduction, waste management plans must address a variety of issues. They must define goals, establish programs, and set up an evaluation mechanism.

To begin with, the plan must quantify a baseline of waste generated. A goal, including a time frame, is then set against this baseline: for instance, a 5 percent reduction by 1996 from 1991 levels (the baseline). The plan must also specify whether the goal is set on a per capita or total waste generation basis.

Before actual reductions in the amount of garbage can be made, the continuing growth in the generation of garbage must be halted. Thus, a good starting point for setting a goal would be to cap total waste generation within 5 years. This could be followed by an actual 5 percent reduction over the next 5 years. That is, if the baseline year were 1991, the total amount of garbage generated in 1996 would be the same as it was in 1991. By 2001, garbage generation would be 5 percent less than 1991 levels.

To meet an overall goal, it is helpful to target specific materials or categories of materials and set up individual reduction goals for them. Communities need to analyze their waste stream to identify its major components and determine which components provide the greatest opportunity for reduction. Then they can establish programs directly aimed at reducing these major components. For example, on a national basis, nondurables and packaging comprise 60 percent of the waste stream.[6] Efforts could be made to increase reusables (such as government procurement policies that favor nondisposable goods) and decrease packaging (such as tax incentives to manufacturers to reduce packaging).

To ensure that goals are met, plans need to include an evaluation mechanism, including standards for measuring and assessing reductions. Uniform national standards for measuring reduction could help states and municipalities track their own progress as well as compare their progress to that of others. It is also important to have a mechanism for determining whether reductions are due to factors other than the implementation of source reduction strategies, such as economic conditions, population changes, or technological changes, among others.

Finally, as communities learn more about and gain greater experience with source reduction, both goals and programs for achieving them can be reevaluated.

Communities can establish individual recycling goals for each locally available material targeted for collection, as well as source reduction goals for different materials. The overall percentage of materials collected for recycling will vary from community to community, depending on the exact composition of its waste stream and on the materials targeted.

At a minimum, state-of-the-art recycling goals could reasonably include 90 percent of yard waste, 85 percent of beverage containers (glass, metal, and plastic), 85 percent of newsprint, corrugated, and office paper, and 20 percent of all other paper. The state of New Jersey has set these goals,[7] which are likely to be understatements of actual possibilities for the collection of materials, for 1995.

The technologies and infrastructure for processing other components of the waste stream are constantly improving and, as they do, it will become increasingly possible to recycle other materials such as food waste (by composting) and an assortment of plastics. Incinerator planners should keep in mind that, during the lifetime of a waste-to-energy plant, the amount of recyclable material is likely to increase.

The necessity of correct sizing

A general, overriding principle for designing a solid waste incinerator is to determine the correct size for the amount of anticipated waste since combustion is most efficient when an incinerator consistently burns the quantity and quality of garbage it was constructed to burn.

The most important limit to the amount of garbage an incinerator can burn is the amount of heat its boiler can handle, often measured in **British thermal units** (BTU, a unit of heat energy) per hour. Different components of municipal solid waste have different BTU values, so the composition of the waste is as important as its volume. Plastics, for example, have BTU values of approximately 13,000 BTU/pound, while paper and wood have values in the range of 6000 to 7000 BTU/pound, and metals and glass have no BTU value since they do not burn.

If an incinerator is oversized (that is, if there is less garbage available for burning than the plant was originally designed to take), then it may operate less than full time. Each start-up and shutdown causes unsteady burning conditions, resulting in reduced overall efficiency. Such unsteady conditions tend to increase generation of products of incomplete combustion and particulates. Building an oversized plant also has economic implications. Burning less than full time decreases the amount of energy produced and sold by the plant. More importantly, a plant that is oversized for the amount of waste available to burn has higher per ton disposal costs. (Building incinerators with separate units that could be operated independently so that the whole plant would not have to be used if the amount of waste decreased could add flexibility and lessen some of the environmental impacts, but would not alleviate the economic problems.)

If an incinerator is undersized (that is, if there is more garbage to be burned than originally planned), too much garbage may be loaded into the furnace. Overloading an incinerator can result in increased generation of products of incomplete combustion, as well as an increase in the volume of unburned matter and ash. An undersized incinerator that is not overloaded will necessitate additional expenditures of alternative methods of waste disposal or recycling.

Measuring and categorizing the waste stream

Waste composition studies to determine the amount and composition of the waste stream, including identifying materials that can be targeted for source reduction or removed for recycling and toxic materials or pollutant precursors that should also be removed, are vital for developing an integrated solid waste management strategy. They are also essential for determining the appropriate size for the plant.

The best method for determining the amount of garbage being generated is to obtain actual waste data just prior to design and sizing, as is done for some Japanese incinerators. Waste composition studies should ideally sample enough wastes from different neighborhoods at different times of the week and year to constitute a representative sample. Within a given area, the amount and type of garbage will vary throughout the year: more yard waste and beverage containers in the summer than in the winter, for example. Most of these seasonal changes occur in residential waste which, on a national level, comprises approximately two-thirds of the municipal solid waste stream; commercial waste makes up the remaining third.[8] (Recent estimates by the Fredonia Group, a Cleveland-based research organization, project residential waste increasing to 72 percent of the waste stream in 1995.[9]) The quantity and components of processible commercial waste must also be identified.

Some communities use average waste composition data from other towns or cities to estimate their waste composition. This method can be misleading, however, since the composition of municipal solid waste changes not only from place to place but also over time; plastics use, for example, has skyrocketed in the past 10 to 15 years.

Information about projected population growth and future trends in the volume and composition of the waste stream is just as critical as current waste data, especially in this time of changing solid waste management methods. It is vital for communities to plan for increased source reduction and recycling before an incinerator is built, since incinerators are typically designed for at least a 20-year lifetime and since recyclables may constitute large percentages of the waste stream. Further, incinerator financial arrangements often include long-term (15 to 30 years) contracts for the quantity of garbage to be delivered to the plant or for the quantity of energy to be sold that may preclude a community from undertaking aggressive source reduction or recycling later on. Knowing what potentially recyclable materials are in the waste stream, and in what quantities, is essential.

The waste composition study should, optimally, be designed to plan both source reduction and recycling actions. Two separate sorts would be necessary to accomplish this. To evaluate source reduction possibilities, the amount of the waste stream made up of durables, nondurables, and packaging must be determined. To plan recycling, the quantity of each recyclable material in the waste stream must be determined. It is preferable, for example, to separately identify white office paper, colored stock, corrugated cardboard, computer paper and cards, magazines, newspapers, and so on, rather than the single category "paper." When marketed, these resource streams are more valuable when separated into specific categories than when they are commingled. Although only some of the potentially recyclable materials from the waste stream can currently be marketed, varying with the area of the country, markets are likely to continue to develop.

It is also desirable to know the amount and source of toxic constituents and pollutant precursors in the garbage to be burned. If not destroyed during burning, these will either be released into the air or accumulate in the ash residue. Sampling for specific products, such as batteries, electronics, and chlorinated plastics, in which pollutant precursors are known or suspected, is important. Subsequent elemental analysis can help to verify which pollutant precursors are present, in what concentrations, and in which products. In this way, intelligent choices can be made about reduction and separation programs.

Ideally, products containing toxic materials should be reformulated with nontoxic substitutes. If this is not possible, they should be separated from the waste stream and not burned at all, as is the goal in Japan. For instance, improved recycling efforts could help remove batteries from the waste stream prior to incineration. Knowledge about what substances are present in what quantities, and about how this might change over time, could also facilitate government and industry efforts to target legislation or regulation at items that could be manufactured with nontoxic substitutes. Cadmium, for example, which is commonly used for coloring plastics and inks, is an especially troublesome and toxic pollutant; its use as a pigment, stabilizer, or coating is prohibited in Sweden.

Finally, plant designers need information about the composition of the waste stream to determine the optimal physical design of the plant. (The factors enabling the cleanest possible furnace operations are discussed below and summarized in Tables 3-2 and 3-3.) As mentioned earlier, different materials generate different amounts of heat energy when burned, and knowing the anticipated overall BTU value is critical to planning boiler capacity and furnace structure.

Processing Wastes Before Burning

A state-of-the-art solid waste management strategy should design for source separation and/or mechanical separation to the greatest extent possible to preserve natural resources, improve incineration efficiency, and minimize emissions and ash quantity and toxicity. The percentage of the solid waste stream that could be separated and not incinerated varies from place to place, depending on the amount of noncombustible material, toxics, and pollutant precursors in the waste. In 1987 in Japan, as much as 50 percent of the original waste stream was recycled. Of the remaining 50 percent, approximately one-third was removed prior to incineration (e.g., noncombustibles, hazardous materials). Thus, overall in Japan, about 33 percent of the waste stream was incinerated and about 17 percent was landfilled.[10]

Many materials found in municipal solid waste are unsuitable for incineration. Highly flammable or explosive materials, when exposed to furnace conditions (and sometimes even mechanical sorting and processing equipment), can cause explosions, endangering worker safety. Burning wastes that contain pollutant precursors such as heavy metals (concentrated, for example, in batteries, electronics, and appliances), chlorine (a component of many plastics), or organic – carbon-based – chemicals (found, for example, in household solvents) can cause higher levels of emissions of toxic substances. The presence of noncombustible materials in the incinerator, such as cans and glass, reduces combustion efficiency, leaves more unburned material for disposal, causes more wear and tear on grates, and coats boiler tubes with partially melted fly ash. Processing prior to incineration is vital to remove as much of these materials as possible from the waste to be burned.

The extent to which wastes are currently processed before incineration varies extensively and is far from the state of the art. Most mass burn plants limit on-site processing to the removal of hazardous or explosive materials and bulky items. Refuse-derived fuel plants, at a minimum, also remove iron-containing **(ferrous)** metals with magnets, and may remove other noncombustibles, such as glass and aluminum, mechanically and by hand; sometimes plastics and other wastes are also removed.

Municipalities can do better now by establishing more extensive **source separation** requirements for garbage at the point of generation (home or business), with curbside collection of, or drop-offs for, sorted materials. Alternatively, mechanical or semimechanical separation can take place at an intermediate processing center where mixed garbage is sorted into different resource categories for sale to intermediate markets (and the remaining materials are shipped to an incinerator or landfill).

Waste processing prior to incineration is very beneficial from an environmental standpoint. Depending on what materials are removed, it:

1. improves the homogeneity of the wastes burned, usually improving combustion efficiency and thereby reducing emissions of products of incomplete combustion;
2. reduces the quantity of pollutant precursor input, and thus the toxic substances in air emissions and leachate from ash (for example, removing batteries that contain mercury should reduce the amount of mercury released into the atmosphere or accumulated in ash);
3. reduces the quantity of ash residue generated; and
4. allows for the recovery and reuse/recycling of resources that would otherwise be wasted, at the same time preserving virgin materials and energy resources.

Both source separation programs and mechanical sorting equipment are now employed in many cities and facilities around the country, with increasing success and efficiency as experience is gained.

Pollution prevention through waste presorting

One of the major benefits of waste presorting is the attendant removal of solid waste items known to contain pollutant precursors — elements or compounds such as heavy metals, chlorine, sulfur, nitrogen, and fluorine that, when burned in a municipal solid waste incinerator, produce emissions such as hydrogen chloride, oxides of nitrogen, sulfur dioxide, mercury and other vaporized metals, and chlorinated organics.

Studies have demonstrated considerable reduction in pollutant production when presorting occurs. For example, studies conducted for the United States Department of Energy by National Recovery Technology demonstrated that presorting of noncombustibles, including glass, ceramics, and metallic wastes (such as aluminum and ferrous cans and batteries), prior to incineration in three types of mass burn plants, brought about a reduction of as much as 50 to 70 percent in the emissions of lead, chromium, and cadmium. In one of the three plants (in Gallatin, Tennessee), a 75 percent reduction in emissions of products of incomplete combustion and a 63 percent reduction in carbon monoxide emissions (another combustion-related parameter) were seen. Ash production rates as a whole were 45 percent less than when unsorted municipal solid waste was burned. The amount of heavy metals in the ash, available for leaching, also decreased (48 percent for lead, 57 percent for silver, and 13 percent for cadmium).[11]

Data also seem to indicate that several metals (for example, copper, cadmium, zinc, antimony, chromium, and lead) serve as catalysts in the secondary formation of dioxins at temperatures typical of the back end of incinerators,[12] so the removal of these metals prior to combustion should reduce the amount of dioxins formed.

A study of the Pittsfield, Massachusetts incinerator indicated a positive linear relationship between chlorine content of waste (based on the relative presence of polyvinyl chloride [PVC], a common consumer plastic used in such products as meat wrapping, bottles for edible oils, and upholstery) and hydrogen chloride emissions; there is disagreement about whether the data also indicated an apparent relationship between chlorine input and furan emissions.[13] Tests conducted by the Environmental Protection Agency indicated that burning equal mixtures of PVC and polyethylene (another common consumer plastic, used in such materials as soft drink bottles and shrink wrap), under test conditions generally considered optimal for combustion, produced a large number of chlorinated organic compounds, including dioxins, furans, and dioxin/furan precursors.[14] While other factors (such as temperature and the presence of chlorinated organic compounds in other materials) also affect the production of dioxins and furans, reducing the amount of these plastics being burned, as well as designing incinerators to assure more efficient combustion, should help reduce emissions of these pollutants under real, as well as test, conditions.

Yard wastes and food wastes account for 25 percent of the waste stream by weight, with local and seasonal variations, and have a high nitrogen content. Since nitrogen in the fuel contributes most to emissions of oxides of nitrogen,[15] separation and composting of these wastes help reduce the emissions of oxides of nitrogen from incinerators. A seasonal comparison of emissions of oxides of nitrogen from several plants of similar technology, prepared by researchers at Ogden Martin (an incinerator company), showed approximately 20 percent less emissions in the winter when there is no production of yard waste than in the spring and summer when collections of yard wastes reach their peak.[16]

Wastes accepted and prohibited by mass burn incinerators

A state-of-the-art solid waste management system would specify exactly which wastes could be burned (based on combustibility and content of toxic materials and pollutant precursors) and would ensure that prohibited materials were detected and removed from the waste. From an environmental perspective, these prohibited wastes would include, at a minimum, recyclables (such as glass and newspaper), compostables (such as yard and food waste), noncombustibles (such as metals), and materials containing hazardous substances (such as batteries).

In reality, however, although most incinerators specify which wastes they will or will not process, there are no uniform standards. Thus, the lists of permitted and prohibited wastes vary from plant to plant, as do the procedures for detecting and excluding prohibited wastes. Table 3-1 is a compilation of materials that have been prohibited at a variety of different municipal solid waste incinerators. Such specifications are stated in contracts between operators and municipalities or in municipal regulations. However, in practice, many of the prohibited materials are burned because plant operators do not prevent them from entering either the plant or the furnace.

Screening at mass burn incinerators

Ideally, unacceptable wastes should be separated, and recycled if possible, before they reach the incinerator, through source separation programs in homes, businesses, and institutions. Those that do end up at the burn plant must be identified and removed. Screening for unacceptable wastes is best accomplished by designing the screening

Table 3-1 Materials Prohibited by a Variety of Mass Burn Plants

Bulky wastes (e.g., furniture) (may be acceptable if reduced in size)
Noncombustible wastes (not including glass bottles, cans, etc.)
Explosives
Tree stumps and large branches (may be acceptable if reduced in size)
Large household appliances (e.g., stoves, refrigerators, washing machines)
Vehicles and major parts (e.g., transmissions, rear ends, springs, fenders)
Marine vessels and major parts
Large machinery or equipment
Construction/demolition debris
Tires
Lead acid and other batteries
Ashes
Foundry sand
Cesspool and sewage sludge
Tannery waste
Water treatment residues
Cleaning fluids
Crank case and other mechanical oils
Automotive waste oil
Paints
Acids
Caustics
Poisons
Drugs
Regulated hospital and medical wastes
Infectious waste
Dead animals
Radioactive waste

area so that waste can be closely inspected and removed if the need arises. Rejected items are usually buried in a landfill. It is essential to employ a sufficient number of workers trained to spot unwanted waste. In a state-of-the-art facility, plant workers would also keep track of the different carting companies that deliver waste to the facility, since some may exhibit a pattern of bringing prohibited materials.

Some preliminary viewing of the waste is desirable when incoming garbage trucks are weighed, but most screening at mass burn plants takes place at the tipping floor or pit. Where the waste is dumped directly into a pit (often the case with large plants), there is little opportunity for the screening and removal of unwanted waste items. **Tipping floors, which resemble large warehouse floors, are better suited for visual inspection and removal of unwanted items.** Further, state-of-the-art screening would include opening garbage bags on the tipping floor, since workers cannot identify unwanted items inside such bags. Figure 3-4 shows garbage stored in a pit at the Baltimore refuse-derived fuel facility.

Figure 3-4

Garbage storage in a pit at Baltimore refuse-derived fuel processing plant (not in INFORM's study). (Photo: David Saphire)

Radioactivity sensors are beginning to be used as a screening device. In 1989, for example, the Bridgeport, Connecticut, mass burn plant returned wastes to a New Haven hospital after such a sensor detected radioactivity.

Waste mixing and drying

Proper mixing and drying of municipal solid waste prior to burning makes the waste more homogeneous and thus permits more even burning with fewer temperature variations, resulting in fewer unwanted emissions. Unstable combustion conditions in the furnace are frequently caused by the constantly varying heat (BTU value) and moisture content of heterogeneous wastes (for example, wet leaves that do not burn well, immediately followed by dry paper, or vice versa).

Mixing can be accomplished by crane operators where garbage is stored in a pit or by front-end loader operators where there is a tipping floor. In refuse-derived fuel plants, mixing occurs as the waste is processed into a more homogeneous fuel. The usual method of drying refuse in a mass burn plant is by exposing it to heat as it travels along a conveyor in the furnace, usually as the first stage of a three-part grate system.

Mixing, drying, and good housekeeping also help retard putrefaction of incoming wastes and the resultant odors. Odors are contained in most modern plants by drawing furnace air into the furnace from a point over the garbage storage area (negative pressure), by avoiding taking in more garbage than can be processed in 24 hours, by periodically disinfecting the facility, and by being careful with opening and closing doors and other openings in the storage area.

Screening and processing at refuse-derived fuel plants

In refuse-derived fuel plants, garbage is sent, usually by conveyor, into a facility separate from the incinerator for preprocessing and sorting. Though plant designs differ, refuse-derived fuel is generally obtained by processing the waste to uniform size and removing metals, glass, and other noncombustible materials. The refuse-derived fuel that remains after processing is highly combustible and can be used as is (a fluffy

Figure 3-5

Magnetic separation equipment at resource-derived fuel processing plant in Baltimore (not in INFORM's study). (Photo: David Saphire)

material) or in pelletized form. The fuel can be burned alone or mixed with another fuel such as wood, peat, coal, sewage sludge, or unprocessed municipal solid waste.

Processing facilities range from simple hand sorting to magnetic separation of ferrous metals (such as cans, steel, steel-clad batteries) to intricate systems in which different devices sort municipal solid waste into many resource streams (for example, glass, aluminum, plastics, and organic materials). (Figure 3-5 shows a magnetic separator at the resource-derived fuel processing facility in Baltimore.) Corrugated paper is usually removed manually. In some facilities, the mechanically separated resources are refined into more saleable commodities (coarse glass to finely ground glass, for example). However, refuse-derived fuel processing that shreds the waste often greatly diminishes the quality of recyclable materials that are more valuable in larger form. **Materials recovery facilities,** or MRF's, on the other hand, also engage in waste processing, but preserve the integrity of the separated materials. In MRF's, paper, plastics, glass, and other potential recyclable products are removed for sale.

Resource-derived fuel processing plants use a variety of shredding and pulverizing devices to reduce the size of the waste, ultimately to particles of 0.25 to 2 inches. The waste is then separated into resource streams and the remaining refuse-derived fuel is stored, transported, and burned as required. Because shredding and pulverizing municipal solid waste produces heat and occasional explosions, refuse-derived fuel processing plants usually require fire suppression systems and explosion chambers for the shredders. They also reject more wastes than mass burn plants since they require more careful inspection of wastes to avoid problems with processing equipment, thus requiring rejected materials to be sent to landfills.

Loading Wastes into the Furnace

The waste feed system introduces garbage into the incinerator furnace from the tipping floor or pit (or, in the case of a refuse-derived fuel plant, from the preprocessing facility). Of the two main types of waste feed system in use today, **continuous loading**

contributes to more efficient combustion than batch loading because it allows for a more even flow of fuel.

In **batch loading,** the waste is introduced into the furnace by a front-end loader truck that shoves the garbage, in discrete batches, into the furnace. With **continuous loading,** a crane deposits waste, a few tons at a time, into the top of an inclined chute. The garbage moves down the chute onto the drying zone of a moving grate, allowing for more continuous introduction of waste into the furnace. Refuse-derived fuel is typically continuously fed into the furnace.

The batch method may adversely affect combustion since each load of garbage pushed into the incinerator causes a temporary overload, depleting available oxygen and creating poor combustion conditions. The continuous loading method enhances combustion efficiency because a more nearly constant volume of garbage passes through the furnace, allowing more nearly steady-state conditions.

While batch loading is predominantly used in plants with tipping floors, and continuous loading in plants with pits, **it is possible to design a plant with both a tipping floor and a continuous loading system to allow for both good screening and continuous operation.** (The Claremont plant has such an arrangement.)

The concept of designing for continuous feed applies to plant operations as a whole: the more often a plant is started up and shut down, the more uneven the combustion and the greater the potential for unwanted emissions.

Combustion

Efficient and even combustion is one of the key factors in minimizing the environmental impact of waste-to-energy incinerators, reducing both the amount of unburned materials in the ash produced, and the amount of air emissions. It depends largely on the design of the furnace and the operating practices used.

The combustion process

From the charging chute, wastes fall onto a series of grates in the furnace in a single-chamber mass burn plant. Air is injected upward through holes in the grates to control combustion. The grates are generally movable. Older systems simply transport the garbage through the furnace. Newer, improved grate systems are designed to also agitate the wastes (by vibrating, rocking, or other methods), thereby promoting combustion by allowing for improved mixing with air. These systems can vary grate speed so that the heat energy produced by the refuse being burned stays at a reasonably constant level and oxygen requirements do not fluctuate excessively.

As garbage is fed onto the grate system, it encounters the high temperatures of the furnace, is heated and dried, and subsequently ignites. In the **primary combustion phase,** burning garbage is then transformed into bottom ash (by-products of noncombustible items in the waste stream) and **char** (unburned carbon-containing materials that remain in the ash). Volatile gases and incompletely burned carbon-based compounds (products of incomplete combustion such as carbon monoxide, dioxins, and furans) are also generated.

When garbage is first exposed to the flames, control of the quantity and temperature of the air injected into the system (primary, or underfire air) is central to promoting efficient combustion. Too much **excess air** (the amount of air over and above that required for combustion itself) cools combustion, and causes more particulates to be entrained or lifted off the grate. Too little air reduces destruction of the products of incomplete combustion. Thus, it is very important to separately and continuously control the placement, velocity, and amount of air going to different parts of the grate area, providing sufficient but not excessive air. Injection of heated air can also help maintain steady-state combustion conditions.

The **secondary combustion phase** takes place as the gases formed in the primary phase rise above the grate and are themselves burned. The **"three T's" — temperature, turbulence, and time — are the factors involved in maximizing combustion efficiency at this stage.**

For most mass burn and refuse-derived fuel incinerators,

1. temperature should fall within the 1800-2000°F range and should be uniform across the furnace at the mixing level;
2. turbulence, or adequate mixing of the combustion gases with oxygen, is accomplished by the injection of air above the grate (**secondary,** or **overfire air**);
3. the time during which gases should be retained at the above conditions, or **residence time,** should be at least 1-2 seconds.

The three T's are critical for all incinerators, but the specific requirements differ. For fluidized bed combustors (another type of furnace), for example, typical conditions are temperatures of 1500-1700°F and residence times may be up to 5 seconds.

Furnace designs

Controlling combustion efficiency by controlling air in both the primary and secondary phases, as well as temperature, turbulence, and time in the secondary phase, depends largely on the design of the furnace and the operating practices used. Modern plant designs achieve superior combustion efficiency by utilizing newer features such as automated combustion controls and physical structures that allow the gases greater opportunity to mix with oxygen in the higher-temperature region of the furnace.

Furnaces operating today vary considerably. Older ones generally do not achieve as efficient combustion as the newer ones. Several basic furnace designs are currently in use (single-, dual-, and multiple-chambered; rotary, and fluidized bed), and several modifications exist within these basic categories.

Theory and experimental tests suggest that certain designs (such as dual-chambered versus single-chambered furnaces and fluidized bed combustors) should allow more efficient combustion with less production of pollutants. In actual practice, many factors other than furnace design, including other plant design features and operational practices, play important roles in determining the environmental performance of a garbage incinerator. When examining the 15 study plants, as will be seen in Chapter 4, INFORM did not evaluate combustion efficiency alone.

Single-chambered furnaces. For large-scale incinerators, single-chambered furnaces with provision for underfire and overfire air are the most common design configuration in use today. They are usually employed in mass burn facilities processing 250 or more tons per day. (The furnace shown schematically in Figure 3-2 is single-chambered.)

Furnace exteriors must be cooled to protect surrounding structures; two main systems exist. With the older method, **refractory walls,** ceramic and stone walls surrounding the furnace reflect heat back into the fire. Water-cooled furnaces, on the other hand, use **water walls:** water-containing pipes in the walls absorb the heat from the furnace and transmit the heat to the contiguous boiler. Small furnaces tend to have refractory walls, while larger ones often have water walls.

To penetrate the waste on the grate, underfire air is sometimes injected at high speeds, leading to entrainment of particulates upward off the grates into flue gases. Temperatures higher than 2000°F, intended to ensure the destruction of products of incomplete combustion, can contribute to production of oxides of nitrogen.

The design of the interior of the furnace and the quality of combustion controls, which vary considerably from plant to plant, can affect combustion efficiency. For example, carefully placed protrusions from the furnace wall, called **arches** and **bullnoses,** can redirect the flow of air from the grate, guiding it into turbulent eddies (or air currents) within the furnace. Eddy currents aid in maximizing turbulence during secondary combustion of the gases. Optimal furnace design also takes into account the furnace volume needed to allow sufficient quantities of overfire air to be injected and sufficient mixing to occur, for the type of wastes to be burned.

Additionally, **auxiliary burners** can prevent major upsets (reductions in combustion efficiency) usually due to one or more of several factors: start-up or shut-down; large changes in the moisture content, heat content, or quantity of incoming refuse; and maladjustment of the air distribution system. An auxiliary burner is a device that burns another, more uniform fuel (such as natural gas or oil) during start-up, shut-down, or upsets, usually when temperature values fall below 1500-1800°F, thereby stabilizing combustion by maintaining a minimum furnace temperature.

Other important features of better single-chambered furnaces are good design of the air distribution system and automatic combustion controls. Early mass burn incinerators had simple underfire and overfire air dampers manually controlled by operators, often with no mechanical devices to draw in air. Frequently, in such older incinerators, holes developed in the furnace walls, allowing air to leak in and degrade combustion efficiency. The newer and better large-scale incinerators direct air to flow in some zones and not others, and at different velocities at different times. They also employ automatic combustion controls that continually regulate the air flow using a computerized system that interprets data provided by temperature and oxygen probes.

The main purpose of these air distribution and automatic combustion control systems is to achieve more stable combustion. However, they may also permit some reductions in particulate entrainment if underfire air can be held to a minimum and in formation of oxides of nitrogen if recirculated air (relatively oxygen-starved gas) can be used.

Dual-chambered furnaces. Dual-chambered furnaces are also used in mass burn plants but usually are designed to burn a smaller quantity of garbage than single-chambered ones. As the name implies, they consist of two chambers: a primary chamber in which primary combustion of solid material takes place and a separate secondary chamber for secondary combustion of gases. Figure 3-6 is a cutaway diagram of a dual-chambered furnace.

The primary chamber typically operates with less air and at a lower temperature than single-chambered furnaces. Additional combustion air is injected into the secondary

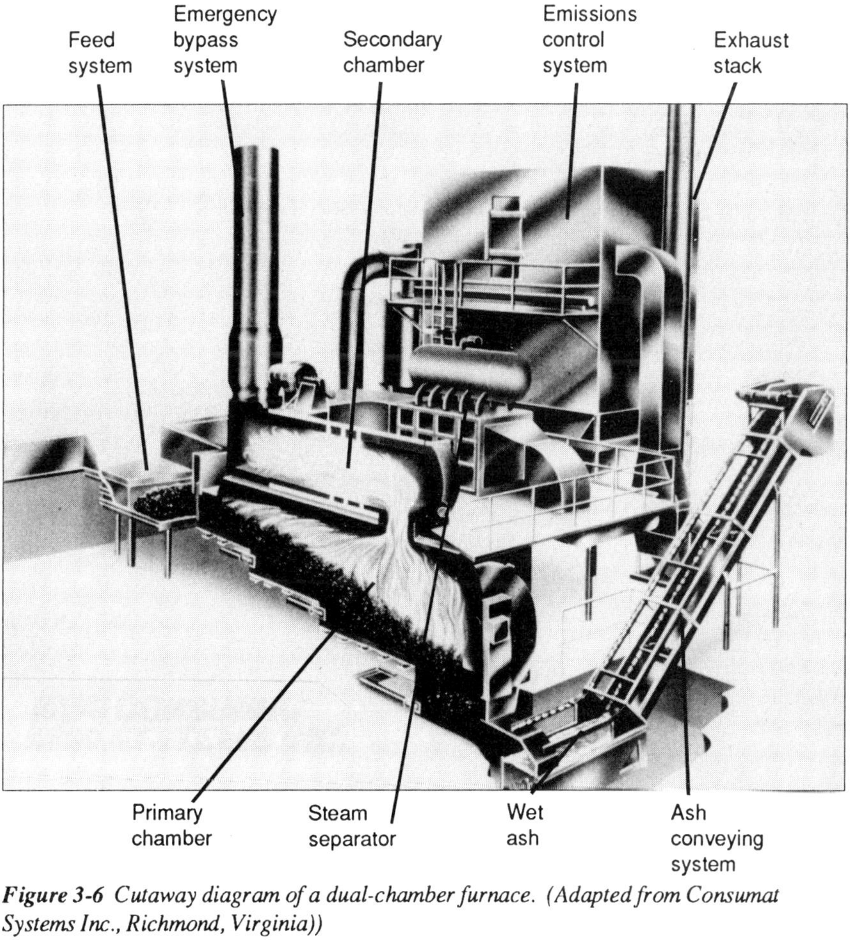

***Figure 3-6** Cutaway diagram of a dual-chamber furnace. (Adapted from Consumat Systems Inc., Richmond, Virginia))*

chamber, which is operated at a higher temperature comparable to that of single-chambered furnaces, to promote the burning of combustion gases. This design is intended to minimize particulate entrainment and to enhance complete combustion of products of incomplete combustion in the secondary chamber. It also usually results in lower formation of oxides of nitrogen in the primary chamber. As with single-chambered furnaces, air distribution systems and automatic combustion controls can be used to improve overall combustion efficiency.

Multiple-chambered furnaces. Furnaces with four chambers have been used for medical waste incinerators for some time. This technology is available now, and could be used, but has not been so far, for garbage incinerators. The third and fourth chambers allow additional combustion to take place, thus ensuring more complete destruction of waste. Recent test data from four-chambered hospital waste incinerators suggest that such multiple-chambered burn systems produce fewer emissions and reduce the formation of toxic forms of dioxins and furans during the combustion process, when compared to single- and dual-chambered furnaces.[17, 18]

Rotary combustors. Rotary combustors are not yet in widespread use for municipal solid waste incinerators (in this study, they are found in the Pascagoula and Tampa plants), although they are often used for burning hazardous waste. A common form of rotary combustor consists of a large, slightly inclined, rotating cylindrical furnace. Waste is fed into the higher end and tumbles slowly towards the bottom. This action continually mixes the waste on the grate and exposes new surface areas to the heat and air, a method designed to aid in improving combustion efficiency. Recently, water-cooled rotary combustors have been developed. These systems result in somewhat lower flame temperatures which, in combination with lower excess air, decrease the potential for formation of oxides of nitrogen.

Fluidized bed combustors. Although fluidized bed technology does not have nearly as extensive a track record as single- and dual-chambered furnaces, it is a promising new system that has the potential to supplant other major furnace configurations as a preferred choice as operational experience with it grows. No plants in this study employed this design, but municipal solid waste incinerators with fluidized bed combustors currently exist in Duluth (Minnesota), Japan, Norway, and Sweden. Existing data demonstrate comparatively (and in some cases, extremely) low emissions of dioxins, oxides of nitrogen, hydrogen chloride, and sulfur dioxide.[19, 20]

In this design, refuse-derived fuel is injected into a loose bed of sand and limestone particles that are in a constant state of turbulence. Air passes through this turbulent bed, where it reacts with the heated refuse-derived fuel and its combustion products. The sand and limestone distribute heat evenly throughout the furnace, thus enhancing combustion efficiency at somewhat lower temperatures (1500-1700°F) than typical mass burn incinerators, reducing the potential for formation of both products of incomplete combustion and oxides of nitrogen. Additionally, the limestone tends to neutralize acids, thus reducing emissions of acid gases. Lower upward velocities out of the combustion chamber and recirculator reduce particle entrainment.

Table 3-2 Summary of Key Factors in Enhancing Combustion Efficiency

Problem	Corrective Techniques
Variation in heat (BTU) content and moisture of wastes	• Presorting garbage before burning to remove noncombustibles • Mixing presorted garbage before loading into furnace to improve homogeneity • Continuous loading of garbage into furnace • Drying refuse prior to burning • Grate designs to promote mixing of garbage (reciprocating, rocking, etc.)
Low (below 1500°F) or fluctuating temperature	• Plug holes in furnace • Automatic combustion controls • Auxiliary burners • Grate design to permit garbage mixing
Uneven temperature across the furnace	• Air distribution systems (correct underfire and overfire quantity and direction) • Auxiliary burners • Plug holes in furnace
Insufficient mixing of combustion gases (including products of incomplete combustion) with oxygen (turbulence)	• Air distribution systems • Arches and bullnoses • Optimal quantity and distribution of air to furnace during the secondary phase of combustion
Insufficient time for gases to burn completely (residence time)	• Arches and bullnoses • Reduce combustion air

Maximizing combustion efficiency

Table 3-2 summarizes the major furnace conditions that result in incomplete combustion and techniques for promoting greater combustion efficiency. In combination, all these techniques lead to more complete combustion.

Minimizing pollutant formation in the furnace

Table 3-3 summarizes the key furnace conditions involved in minimizing production of the main pollutant types (particulates, heavy metals, products of incomplete

Table 3-3 Summary of Key Factors in Minimizing Pollutant Production in Furnaces*

Pollutant	Method of Formation
Particulates	Noncombustible material lifted off grate into flue gas
Heavy metals	Metals volatilize at high temperatures; high upward velocity lifts them off grate (mercury is an exception).
Products of incomplete combustion	Incomplete combustion of varied waste stream containing pollutant precursors
Acid gases	Pollutant precursors in waste stream (e.g., chlorine, fluorine, and sulfur) react with hydrogen or oxygen in the furnace
Oxides of nitrogen	Nitrogen in fuel and furnace air reacts with oxygen at temperatures above 1600°F.

* *Note that an essential technique for minimizing pollutant formation is removal of materials containing pollutant precursors before they enter the furnace. For example, as discussed in the section on pollutant prevention through waste presorting, removing aluminum and ferrous cans and batteries reduces heavy metal emissions, as well as secondary formation of dioxin/furans, and removing food and yard wastes reduces production of oxides of nitrogen.*

Optimal Conditions for Minimizing Formation
• Less air introduced during primary combustion reduces potential for light unburned material to be lifted off the grate. • Arches and bullnoses reduce upward velocity of gases. • Multiple combustion chambers.
• Lower temperatures in primary combustion reduce volatilization. • Less air injected during primary combustion reduces potential for unburned metallic material to be lifted off the grate. • Multiple combustion chambers.
• Maximizing combustion efficiency minimizes formation of products of incomplete combustion. • High temperatures and adequate oxygen during secondary combustion destroy products of incomplete combustion formed during primary combustion.
• Less air introduced during primary combustion reduces oxygen available for reaction with sulfur, as does flue gas recirculation. • Fluidized bed furnaces contain limestone, which tends to neutralize acids. Injection of lime or other alkaline agents into furnace can have similar effect.
• Less air introduced during primary combustion reduces oxygen available for reaction with nitrogen. • Lower temperature in primary combustion reduces potential for formation of oxides of nitrogen. • Flue gas recirculation systems feed gases that have exited the boiler (and thus contain less oxygen) back into the furnace.

combustion, acid gases and oxides of nitrogen). It must be stressed, however, that fuel cleaning (sorting and reduction), design modifications, operational practices, and emissions control devices all have significant further effects on ultimate emissions levels.

Heat Recovery

In waste-to-energy plants, heat generated during the incineration process is used to heat water to make steam. In single-chambered, mass burn incinerators, for example, as gases flow upwards in the furnace, some of the heat generated by the burning garbage is transferred through the furnace wall. When the combustion gases exit the furnace and enter the boiler, their temperature is around 1500°F. At this point, energy is recovered as the heat from the gases converts the water in the boiler tubes to steam that may be used to generate electricity in a turbine. Some of the steam or electricity may be used on-site, with the balance exported. After most of the heat has been extracted, the flue gases leave the last section of the boiler (the **economizer**) at temperatures around 450°F.

During the passage of gases through the boiler, particulate matter, or fly ash, is deposited on the walls and tubes of the boiler. This may collect for long periods before being cleaned, and precursors such as chlorobenzene and chlorophenol in the gases can react on the surface of the soot particles at this temperature to form dioxins and furans. It is believed that the temperature conditions in the last part of the boiler aid in this secondary formation of dioxins. Increased soot removal frequency and innovative cleaning techniques may help minimize this secondary formation of dioxins and furans. Cleaner tubes have fewer particulates on which dioxins and furans can form and allow more heat to be transferred away from the flue gases, thus further cooling the gases below the 450°F temperature which is conducive to dioxin/furan formation. Additionally, minimizing the production of precursors in the furnace by maximizing combustion efficiency helps decrease secondary dioxin/furan formation.

Emissions Control

Although incinerator design, operating practices, and fuel cleaning (waste reduction and separation systems) can significantly reduce the amount of pollutants produced in waste-to-energy plants, some pollutants are inevitably generated. Add-on emissions control devices are used to neutralize, condense, or collect these pollutants, once they are generated, to prevent them from being emitted into the air. Most of these are placed at the so-called "back end" of the incinerator, treating the flue gases after they pass out of the boiler.

Different types of pollutants require different control devices: scrubbers and condensers for acid gases; scrubbers and condensers with electrostatic precipitators or fabric filters (baghouses) for mercury, dioxins, and furans; fabric filters or electrostatic precipitators for particulates and other heavy metals; and chemical neutralization systems for oxides of nitrogen. Variation exists within these basic categories: while some devices are more likely than others to achieve very high removal efficiencies, operational factors, such as temperature, also play a key role.

Since incineration is a complex process using a heterogeneous fuel, devices that control emissions of one pollutant may have other effects on other pollutants. It is important to consider these interactions when introducing new control technologies and to identify techniques for minimizing any undesirable side effects.

State-of-the art emissions levels

Table 3-4 lists state-of-the-art emissions levels for different pollutants. These levels were determined by INFORM based on data from several sources:

- emissions limits found in permits for existing United States plants;
- recent test reports from incinerators in the United States, Canada, and Europe;
- regulations and recommendations of environmental regulatory authorities, including the United States Environmental Protection Agency, the Swedish Environmental Protection Board, and Environment Canada;
- technical papers from professional conferences, proceedings, and journals.

All of these levels have been achieved in practice, with regularity, utilizing currently available technology, and no attempt has been made to correlate emissions levels with any possible health impacts. The discussion of different control devices refers to these state-of-the-art levels, as does the analysis of the 15 study plants in Chapter 4. In many cases, as Chapter 6 illustrates, federal and state regulations mandate emissions limits that are considerably less stringent than these currently attainable state-of-the-art levels.

Controlling acid gas, mercury, and dioxin/furan emissions

Scrubbers, followed by an efficient particulate control device, have been the state-of-the-art equipment for controlling emissions of such acid gases as hydrogen chloride, hydrogen fluoride, sulfur dioxide, and sulfuric acid, although new techniques are being developed in Europe. While emissions of hydrogen chloride and sulfur dioxide from scrubber-equipped incinerators have been reported to be under 10 ppm, such levels have not yet been achieved with sufficient regularity for INFORM to lower its state-of-the-art levels below 25 and 30 ppm for, respectively, hydrogen chloride and sulfur dioxide. However, they do point out that progress continues to be made in reducing acid gas emissions and that state-of-the-art levels may be lower in the future.

Scrubbers generally use impaction, condensation, and acid/base reactions to capture the acid gases in the flue gas. Since greater removal efficiencies usually accompany greater condensation, devices that lower gas temperatures and thus increase condensation can enhance scrubber effectiveness. The lower temperatures may also allow mercury, dioxins, and furans to condense so they can subsequently be captured by a particulate control device. Figure 3-7 shows a schematic diagram of a scrubber, in this case followed by a fabric filter (baghouse) for particulate control.

Three main types of scrubbers are in use today: wet scrubbers, spray-dry scrubbers, and dry injection scrubbers. The first two also act as condensers, while dry injection

Table 3-4 State-of-the-Art Emissions Levels*

Pollutant/ Emission Level	Sources†
Particulates/ 0.010 grains per dry standard cubic foot (gr/dscf)	• **State permits and regulations:** Permits in Maine, California, and Illinois, and regulations in New York, set limits of 0.010 gr/dscf. • **Some actual levels achieved:** Framingham, Massachusetts, 0.007 gr/dscf; Wurzburg, Germany, 0.004 gr/dscf; Rutland, Vermont, 0.003 gr/dscf.
Dioxins and furans/ 0.10 nanograms per dry normal cubic meter (Eadon toxic equivalent)(ng/dNm^3)	• **Regulations:** Swedish Environmental Protection Board set goal of 0.5-2 ng/Nm^3 for existing incinerators and 0.1 ng/Nm^3 for new plants; the Netherlands, Austria, and Germany set limits of 0.1 ng/Nm^3 for new plants. • **Some actual levels achieved:** below 0.1 ng/Nm^3 in Malmo, Stockholm, and Linkoping, Sweden, and in Quebec City, Canada.
Carbon monoxide/ 50 parts per million (ppm), 4-hour average	• **Recommendations:** US EPA's "Municipal Waste Combustion Study — Report to Congress, June, 1987" and Environment Canada's "Operating and Emissions Guidelines for Municipal Solid Waste Incinerators, October, 1988."
Sulfur dioxide/ 30 parts per million (ppm)	• **Regulations:** US EPA's New Source Performance Standards. • **Plant permits:** Permits in California and Maine set limits of 30 ppm.
Hydrogen chloride/ 25 parts per million (ppm)	• **Regulations:** US EPA's New Source Performance Standards. • **Some actual levels achieved:** 19.5 ppm in Jackson County, Michigan; 1.7 ppm in Rutland, Vermont.
Oxides of nitrogen/ 100 parts per million (ppm)	• **Plant permits:** Permits in Illinois and Maine set limits of 100 ppm. • **Some actual levels achieved:** 90 ppm in Rutland, Vermont; 93 ppm in Long Beach, California; 37 ppm in Bremerhaven, Germany.

* State-of-the-art emissions levels identified by INFORM are standardized to conditions of 7% O_2. Levels described by other sources may be standardized differently. Appendix B, on methodology, describes the standardization techniques used by INFORM.

† These emissions levels are also supported by two technical papers published by INFORM: *Improving Environmental Performance of MSW Incinerators* and *Technologies for Minimizing the Emission of NO_x from MSW Incinerators,* both by Marjorie J. Clarke.

Figure 3-7 Schematic diagram of Commerce, California, incinerator with scrubber, fabric filter (baghouse), and chemical control device for oxides of nitrogen. (Provided courtesy of: Commerce Refuse-to-Energy Authority)

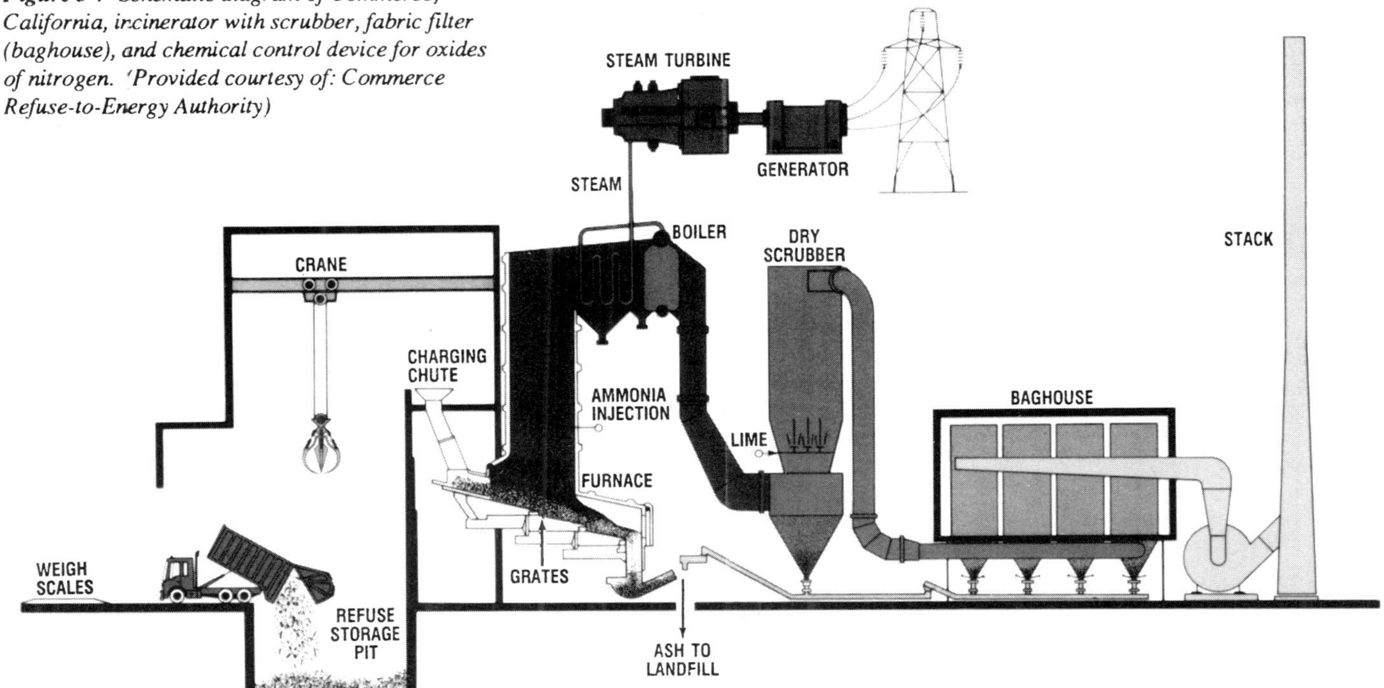

scrubbers require a separate condenser (either a humidifier or a heat exchanger). In all cases, temperature and, for dry and spray-dry scrubbers, the amount of lime (an alkaline substance that neutralizes acids) used are the key factors affecting scrubber effectiveness. In general, to maximize emissions control, the scrubber should be adequately sized, operate at temperatures well below 300°F (and preferably below 270°F), and allow flue gas circulation through the scrubber for at least 10-15 seconds.

Wet scrubbers capture acid gases by collecting the acid gas molecules onto water droplets; sometimes alkaline agents are added in small amounts to aid in the reaction. They have a long operating history in Europe and are considered reliable. Newer designs have been reported to achieve over 99 percent removal of hydrogen chloride and, in some cases, of sulfur dioxide, and over 80 percent removal of dioxin, lead, and mercury.[21] Disadvantages in the past have included added costs for treatment of wastewater produced, corrosion of metal parts, high operating costs, and incompatibility with the fabric filter type of particulate control device. Newer designs have begun to address these concerns.

Spray-dry, or semi-dry, scrubbers operate similarly to wet scrubbers, but the acid gases are captured by impaction of the acid gas molecules onto an alkaline slurry (such as lime). The principal advantage of this type of scrubber over wet scrubbers is that the residue produced (a dry fly ash/lime mixture) is devoid of water and hence easier to treat or dispose of; additionally, the power requirements and corrosion potential are less. Emission tests have demonstrated control efficiencies of 99 percent or better for hydrogen chloride and sulfur dioxide under optimal conditions (temperatures well below 300°F, sufficiently high lime/acid ratios, and sufficiently high gas residence time in the scrubber); dioxins were also considerably reduced.[22]

Dry injection scrubbers inject dry powdered lime or another alkaline agent that reacts with the acid gases in the flue gas. In one research test, removal efficiencies of 99 percent for hydrogen chloride and 96 percent for sulfur dioxide were measured under optimal temperature conditions (230°F); removal efficiencies were dramatically lower at higher temperatures.[23]

In addition to scrubbers, new techniques for reducing acid gas, mercury, and dioxin emissions are being developed and installed on new, existing, and pilot incinerators in Europe. For example, injection of special reagents, such as sodium sulfide or activated carbon, may enhance mercury and dioxin capture, and incorporation of a bed of activated carbon at the end of all of the pollution control equipment may further drive down emissions of oxides of nitrogen as well as acid gases, mercury, and dioxins.

Controlling particulate and heavy metal emissions

Emissions of particulates and heavy metals are best reduced by collecting them in one of two basic types of add-on particulate control devices, fabric filters and electrostatic precipitators. The heavy metals are captured because they condense out of the flue gas onto the particles. These devices are designed to operate at temperatures lower than the approximately 450°F temperature of flue gas leaving the boiler; some operate at temperatures as low as 250°F, beneficial for condensing and collecting acids, volatile metals, and organics. The state-of-the-art level for particulate emissions is 0.010 grains per dry standard cubic foot (gr/dcsf).

Fabric filters (also called **baghouses**) are a **state-of-the-art particulate control technology capable of consistent 99 percent removal efficiency over the range of particle sizes.** They are large structures containing woven bags of fabric that work much like vacuum cleaner filters. (See Figures 3-7 and 3-8.) Flue gas is drawn through the bag filter; particulate matter remains on the fabric and accumulates. Particulates as small as 0.1 microns may be captured. The accumulated particulates, or fly ash, fall into a hopper when the fabric filters are cleaned, and this ash must be disposed of appropriately.

Electrostatic precipitators (ESPs) consist of one or more pairs of electrically charged plates or fields. The particulates in the flue gas are given an electrical charge, forcing them to be drawn out of the gas stream to stick to the oppositely charged plate. The plates are cleaned by rapping and, again, the ash requires disposal. **Electrostatic precipitators with four or more fields are state-of-the-art devices.** The schematic diagram in Figure 3-2 includes an electrostatic precipitator.

A third type of particulate control device is the cyclone, a mechanical device that funnels flue gases into a spiral, creating a centrifugal force that removes larger particles. While cyclones are the most primitive particulate collection device in use, they can improve the efficiency of fabric filters and electrostatic precipitators, when used in combination, by removing larger particles before they reach the other more efficient particulate control devices.

State-of-the-art technology for particulate removal devices has evolved rapidly. Two-field electrostatic precipitators, which were state of the art in the early 1980's,

Figure 3-8

Fabric filter at Poughkeepsie, New York incinerator (not in INFORM's study). (Photo: David Saphire)

have become outmoded, and even three-field electrostatic precipitators have been surpassed by four- and five-field devices and by fabric filters. In the past four to five years, tests have indicated that **state-of-the-art emissions levels (less than 0.010 gr/dscf) are commonly attainable with either fabric filters or four-field electrostatic precipitators.**

Particulate control devices will also collect heavy metals and other pollutants that have condensed out of the flue gas onto particle surfaces when they are placed after a scrubber. Placing a scrubber first also helps lower the temperature of gases entering the fabric filter or electrostatic precipitator. However, wet scrubbers cannot precede fabric filters because the wet particles in the flue gases will clog the filters. Thus, facilities with wet scrubbers, such as the Lakeland plant examined in this study, place the scrubber after the particulate control device.

The smallest particles are the most potentially damaging when inhaled into the lungs, and it is on these smaller particles that dioxins, furans, acid gases, and heavy metals adsorb in greatest quantity. **Thus, a state-of-the-art particulate collector should achieve even lower emissions levels for particulates below 2 microns in diameter.** California and Maine have set emissions limits of 0.008 gr/dcsf for this finer fraction.

Since many heavy metals condense at temperatures of 450°F, both precipitators and fabric filters collect heavy metals that condense onto particulate matter. Effective mercury emission control technology, while evolving, has not yet been fully developed or implemented for municipal solid waste incinerators, but wet scrubbing, condensation, and activated carbon and sodium sulfide technologies show promising results. Minimizing the amount of waste containing heavy metals that enters the incinerator also reduces heavy metal emissions and the presence of such metals in ash.

Controlling emissions of oxides of nitrogen

State-of-the-art control of nitrogen, at the present time, requires both minimizing the formation of oxides of nitrogen in the first place and destroying those that are formed. Strategies for minimizing formation include utilizing appropriate incinerator designs (such as flue gas recirculation and/or dual-chambered furnaces) and operating practices (such as optimal temperatures and amounts of excess air) and incorporating yard/food waste separation into a municipality's solid waste management strategy. Techniques for destroying oxides of nitrogen generally involve injection of chemicals that neutralize them.

Chemical injection control devices use ammonia, urea, or other compounds to react with the oxides of nitrogen created in the furnace and transform them to nitrogen and water vapor. The currently available technologies for neutralization and removal of oxides of nitrogen from flue gases are called selective noncatalytic reduction (SNCR) and selective catalytic reduction (SCR). Both have been demonstrated successfully on municipal solid waste incinerators, and some have reduced emissions of oxides of nitrogen to state-of-the-art levels of 100 ppm or less. The capacity for wet scrubbing and condensation to achieve control of oxides of nitrogen has also been demonstrated.

In the most well-known **selective noncatalytic reduction** technology, Exxon's Thermal De-NO_x, ammonia is injected into the furnace. The data on this system indicate a removal rate on the order of 50-80 percent, depending on the ammonia injection rates, temperature and oxygen conditions in the furnace, and degree of combustion control.[24,25] Figure 3-7 schematically shows the location of the ammonia injection control device in a Thermal De-NO_x system.

Another noncatalytic technology, Fuel Tech's NO_xOUT system, involves injection of aqueous urea and other enhancer chemicals into the boiler at various locations; its developers assert that the use of urea is safer than the storage, transportation, and use of ammonia. This technology has effectively reduced emissions of oxides of nitrogen from boilers burning a variety of fuels, and tests have successfully demonstrated this technology on municipal solid waste incinerators.

Selective catalytic reduction operates by injecting ammonia into flue gas after the gas passes through particulate control devices and before it passes through a catalyst bed consisting of metallic materials in a variety of forms. This system has removal efficiencies similar to Thermal De-NO_x, but uses half the ammonia. Disadvantages include the expense of using scarce metals, the depletion of mineral resources, and the addition of heavy metals from the catalyst bed to the solid waste stream upon disposal (unless separated for recycling).

Wet scrubbers, combined with condensers, can reduce emissions of oxides of nitrogen by dissolving nitrogen dioxide in the scrubber. Tests of this technique at Rutland, Vermont, and Duluth, Minnesota, have shown significant reduction.

Optimal arrangements of emissions control devices

The arrangement of the emissions control devices other than the device for oxides of nitrogen is usually standard: a scrubber and condenser, followed by a particulate collector, followed by an induction fan that sucks the flue gases towards and up the stack. There are two reasons for this.

First, fabric filters cannot operate at the high temperature at which gases exit from the boiler without risk of fire. Thus, placing the scrubber between the boiler and the fabric filter or electrostatic precipitator permits the cooling and often the humidification that can prevent fires. Cooling the gases also plays a role in reducing acid gas, mercury, and dioxin emissions. Second, dioxins and heavy metals are trapped more effectively by the particulate control devices when they have first been condensed out of the flue gas and adsorbed onto the surface of the particulate matter, as happens in a scrubber/condenser system.

An alternative arrangement, common in European plants, involves an electrostatic precipitator followed by a wet scrubber. The electrostatic precipitator is not damaged by the high temperatures, and the wet scrubber cools and condenses gases and captures particulates.

The location of the emissions control devices for oxides of nitrogen depends on the type of technology used, as discussed above, and can be in the furnace or the boiler as well as the "back end" of the plant.

Figure 3-9

Control room at incinerator in Poughkeepsie, New York (not in INFORM's study) (Photo: David Saphire)

Day-to-Day Operations

Established monitoring and maintenance procedures of waste-to-energy plants help ensure the best possible operation from an environmental perspective. Worker training and provisions for worker safety are also key to effective plant operations.

The control room

Control rooms are the operating center of a waste-to-energy plant. There, in a well-run plant, operators are kept up-to-date regarding the plant's operating conditions. Information about temperature, steam flow and pressure, air emissions, and equipment condition is presented on control panels, video screens, and lighted displays. However, there are no standard industry criteria for functions to be monitored or the design of such control panels. (Figure 3-9 shows the control room at an incinerator in Poughkeepsie, New York, not in INFORM's study.) Additionally, in some plants, radio and telephone communication are used; in some, employees move throughout the plant to keep themselves informed of plant conditions. State-of-the-art operations include periodic checking of the accuracy of control panel data, as well as keeping the control room staffed at all times.

Continuous monitors

Continuous process monitors (CPMs) and **continuous emission monitors (CEMs)** track the environmental performance of incinerators at all times so that, in the event of combustion upsets or high emissions of one or more pollutants, corrective measures can be implemented in a timely fashion. These monitors can be connected to alarms that warn plant workers of any combustion, emissions, or other operating condition that requires attention.

State-of-the-art CPMs and CEMs measure nine operating and emissions factors: furnace and flue gas temperature, steam pressure and flow, oxygen, carbon monoxide, sulfur dioxide, oxides of nitrogen, and opacity (a crude

measure for particulates). Continuous monitoring of hydrogen chloride is also possible, but is not yet (although it may soon be) a sufficiently widely accepted technique to be considered a state-of-the-art requirement. By monitoring parameters that indicate combustion efficiency (carbon monoxide, carbon dioxide, oxygen, and furnace temperature), plant operators also obtain indirect indications of levels of products of incomplete combustion; measuring them directly is expensive and does not provide information in time to correct a problem. Continuous monitors must undergo frequent maintenance, including periodic calibration, to ensure their accuracy.

Telemetering, or instantaneous computer transmission of continuous monitoring data to local or state authorities (such as environmental protection departments) can be an excellent method of ensuring adequate, sustained environmental performance, provided that the results are in fact monitored by authorities in environmental protection and public works departments. A valuable technique for providing citizens with information, which has been used in Japan, involves publicly displaying emissions data on monitors outside the garbage-burning plant. Maine is now mandating telemetering for its newer plant permits, and several other states — Maryland, Delaware, and New Hampshire, among them — are considering such a requirement.

Maintenance

Routine troubleshooting and regular maintenance of waste-to-energy plants not only keep the plant's equipment in good operating order, but also help the plant achieve its best environmental performance. Regular maintenance, by keeping operational costs as low as possible and by minimizing unscheduled downtime, is also likely to improve a plant's long-term financial performance.

Worker Training and Safety

Since resource recovery plants are large, complex, mechanical systems that generally operate 24 hours a day (except for repair and maintenance), process a constantly changing waste stream, and are capable of emitting a large variety of air pollutants from the stack and fugitive dusts from ash handling, the training of operators is vital to optimal functioning and environmental performance. In fact, after the design of a plant is optimized, its performance is largely dependent on the quality of operations and maintenance. In order to optimize the operations and maintenance, workers not only must follow a manufacturer's procedures manual, but also need to understand the effects of variation in the composition and moisture of the waste stream on the functioning of the combustion system, the emissions control systems, and environmental performance, and know how to correct upsets and restore steady-state conditions. In a state-of-the-art system, workers are required to have appropriate training for their particular positions.

Types of plant operators

The number and types of workers at resource recovery plants vary to some degree depending on plant size and design. For example, some very small plants (under 200 tons per day) may have just a few workers per shift to undertake all the tasks of running a plant; in contrast, some of the larger plants (over 2000 tons per day) may have two dozen workers or more, with staff specialized for different tasks.

The design of a plant (for example, mass burn or refuse-derived fuel) also makes a difference in the types of operator needed for a plant. In general, all plants have a chief facility operator and shift supervisors. Operator specialties for a mass burn plant can include weigh scale operator, load inspector, traffic coordinator, crane operator, front loader operator, control room operator, emission controls operation and maintenance personnel, ash management personnel, furnace maintenance specialists, and boiler/turbine operation and maintenance specialists, among others. For refuse-derived fuel plants, workers are also needed to operate and maintain the separation devices and conveyors.

State-of-the-art training

State-of-the-art training for upper-level plant operators (chief facility operator, shift supervisor, and control room operator) involves both formal academic and practical education, as well as supervised on-the-job training. At a minimum, certification should require the following components.

- A four-year bachelor's degree in a technical field, such as science or engineering. (A two-year degree for control room operator may be sufficient.) Required courses should include physical and organic chemistry, fluid dynamics, thermodynamics, materials science, combustion theory, environmental engineering, air pollution control, environmental sciences, toxicology, environmental health, environmental policy and environmental law, with further electives in mechanical, civil, chemical, and electrical engineering.
- Specialized practical training in a six-month program that includes work in simulation situations and other types of laboratories, to learn recognition of hazardous/unwanted materials, control room operations, air and ash sampling/analysis, boiler and turbine operations, recognition and avoidance of upset conditions, and continuous monitor operations and calibration, as well as other topics. Attention should also be given to Occupational Safety and Health Act (OSHA) and worker safety issues.
- Successful work experience in a resource recovery plant with on-the-job training and close supervision for at least six months in the position.
- Formal evaluation and testing prior to certification.
- Periodic review of operator skills and regulatory knowledge. To maintain their certification, operators should be required to demonstrate understanding of changing regulations, new technological achievements (whether or not directly applicable to their plant), and changes in requirements for individual plants.

Lower-level employees, such as crane operators, tip floor personnel, and pollution control equipment technicians and mechanics also need formal and on-the-job training before being certified for specific jobs.

Although on-the-job training is normally conducted on the premises of a resource recovery plant, and bachelor's degrees are obtained at colleges and universities, the theoretical and practical training segments should be conducted in training institutes accredited by the regional education accreditation boards. The facilities and equipment needed for specialized instruction in such a program would include, at a minimum, a municipal solid waste combustion simulator, a chemistry lab, a physics lab, and a computer facility.

The faculty in such programs should minimally include environmental engineers (air quality, air pollution control, emissions modeling, risk assessment), environmental scientists (toxicology, environmental impacts), medical doctors (human health effects), electrical and mechanical engineers (boiler, turbines, combustion theory, and so on), process chemists (municipal solid waste composition, air and ash sampling and analysis), and physicists (fluid mechanics, thermodynamics).

Existing and proposed standards

Until recently, there have been no standardized national programs for training or certification of any level of incinerator operator in the United States. Very few states have operator certification requirements.

Methods currently used to train incinerator operators mostly consist of on-the-job training and in-house training programs. Larger vendors (both waste-to-energy companies and suppliers of pollution control equipment) are likely to provide more on-the-job training for plant employees, as well as formal seminar training. However, this education is by no means uniform in length of time, specificity, or adequacy of curriculum, even among plants of similar size and design. Some operators in waste-to-energy plants have previous boiler operating experience outside the municipal solid waste field. Relatively few have previous experience burning garbage.

The lack of standards for worker certification, although not for worker training, is changing. In 1989, the American Society of Mechanical Engineers (ASME) promulgated a national standard for certification of chief facility operators and shift supervisors that was adopted by the United States Environmental Protection Agency in its New Source Performance Standards, released in early 1991.

These standards fall far short of a state-of-the-art training and certification program. They mandate no formal education or training beyond a high school education, apply only to the two highest-level plant workers, and limit recertification requirements to an annual review by employees of plant-specific operations and maintenance manuals that are not developed according to any agreed-upon standards.

Worker safety

Worker safety in waste-to-energy plants is accomplished through the use of hard hats, eye protection masks, and ear protection equipment. Respirators may be used at times, and are especially advised in the ash handling sections of the plants where the potential of inhaling fine particles is great.

Ash and Its Management

The first priority in state-of-the-art ash management is to reduce both the volume and toxicity of the residue left after burning municipal solid waste by removing noncombustible materials and those containing toxic substances from the waste stream before incineration, and by ensuring efficient combustion. As discussed in the "Environmental Impacts" section of this chapter, the ash amounts can range up to 50 percent of the original waste, by weight (and more in poorer or very old incinerators), and the amount of toxic materials in ash has been increasing as more effective air pollution control devices capture more pollutants in the fly ash. As Figure 3-10 shows, when noncombustible materials are included in the incinerator feedstock, many materials larger than ash remain in the ash residue because they do not burn completely.

Once ash has been formed, it must be handled, treated, transported, and disposed of or reused. State-of-the-art management practices are designed to minimize worker and citizen exposure to potentially toxic substances in ash during these processes.

The issue of ash testing

The growing concern about ash toxicity has increased the demand for ash testing as a basis for determining what disposal methods are necessary. However, there is no agreement about which, if any, of the existing testing procedures provides reliable information.

Many plants have been required to test ash using EPA's **Extraction Procedure Toxicity (EP Tox) Test.** This federal test, designed to determine the leachability of metals from solid waste in a landfill, uses acetic acid to dilute the ash. It addresses eight metals: arsenic, barium, cadmium, chromium, lead, mercury, selenium, and silver. The test has been criticized by some vendors, municipalities, and regulators for not mirroring the actual conditions typical of an ash monofill (pH of rainfall responsible for leaching, wide range of particle sizes) and for being unreliable and statistically invalid for use on ash.[26] Starting in early 1991, the EPA is requiring plants to use the **Toxic Characteristic Leaching Procedure (TCLP)** instead of the EP Tox test in an attempt to more accurately replicate actual landfill conditions and to test for additional pollutants that may be present in ash. None of the other test procedures in use (Canada's test, California's Wet Procedure, for example) has been widely accepted as an accurate or reliable indicator of ash leaching potential.

Sampling ash for testing also presents problems. Since ash varies constantly, many samples must be taken to ensure that the samples are representative of the ash. But sampling and analyzing large volumes of ash on a sustained basis is expensive, and there is no agreement about how to take smaller and less frequent, but still representative, samples.

While testing is important, uncertainty in both sampling and analysis procedures is leading regulators to specify handling, treatment, and disposal technologies for all incinerator ash, rather than requiring testing of individual samples to determine the appropriate management methods.

Ash handling and treatment

Separate handling and treatment of fly ash may result in fewer environmental impacts than handling fly and bottom ash combined when the combined ash is disposed of, untreated, in a landfill or ashfill.

(A)

(B)

Figure 3-10

(A) Materials that did not burn after passing through incinerator at Tulsa plant. (Photo: Maarten de Kadt) (B) Metals sorted from ash for recycling at Baltimore plant. (Photo: David Saphire)

In Europe and Japan, fly ash and bottom ash are kept separate so that the more toxic fly ash can be handled and treated with greater care. In the United States, on the other hand, combining the two ash streams has been the more common handling method, with plant operators expecting that diluting the more toxic fly ash with the less toxic bottom ash would reduce the leaching potential of the fly ash to a level that might not concern regulators. It is also expected that, if ash is wet, it will form a cement by reacting with lime from scrubbers.

The practice of combining fly and bottom ash has come under increasing attack from environmental groups in recent years since laboratory tests have revealed that combined ash contains toxic levels of some substances much more frequently than does bottom ash alone.[27] However, two courts have ruled that the federal Resource Conservation and Recovery Act (RCRA) exempts ash from municipal solid waste incinerators from regulations for handling hazardous wastes. The Environmental Defense Fund believes RCRA does allow such ash to be considered legally hazardous, and these decisions are being appealed. The issue of ash classification is discussed further in Chapter 6 on regulations.

Two main categories of ash treatment — both recently developed and still being improved — are fixation or cementation and vitrification. Both technologies are designed to minimize the environmental impacts of ash and enable the reuse of ash in certain situations (cinder blocks, reefs, roads). The long-term effectiveness of these technologies — that is, their ability to immobilize heavy metals to prevent them from reentering the environment over time — is unproven.

In **fixation** or **cementation,** ash is mixed with cement and/or alkaline scrubber materials to create a hard mass with less leaching potential. Cemented ash is being tested in cinderblocks and artificial ocean reefs, and is landfilled in Japan. Its leachate potential is being tested.

With **vitrification,** heated ash is quickly cooled, usually in an in-plant facility, forming an impermeable, glassy product. The American Society of Mechanical Engineers is conducting research to determine the feasibility and environmental impacts of vitrification.

Although these two treatment technologies are still very new, a few incinerators use them. SEMASS, a refuse-derived fuel incinerator in Massachusetts, employs fixation, and a few Japanese plants (for example, the Sokha City facility) have on-site vitrification facilities.

Another new treatment technique involves washing toxic materials out of the ash with hot water and then treating the water to remove soluble toxic materials. This system has been used in Europe, particularly in incinerators with wet scrubbers.[28]

Ash transportation

Minimizing worker and public exposure to ash involves keeping it wet (since dry ash can be easily dispersed) and keeping it contained. Figure 3-11 shows the Tulsa incinerator, where uncontained ash is left in open storage piles inside the plant. Generally, ash is cooled in a water bath, and thus wet ash can be removed to a waiting truck. A closed system of conveyors is preferred to handling ash in the open. Similarly, the trucks carrying the ash to the treatment facility, landfill, or ashfill should be covered and sealed to avoid dry ash blowing around or wet ash leaking from trucks.

Ash disposal

One challenge for ash disposal is minimizing the leaching by rainwater of toxic metals that can contaminate the groundwater and enter the food chain. Preliminary

Figure 3-11 *Uncontained ash inside the Tulsa plant. (Photo: Maarten de Kadt)*

indications are that vitrification followed by disposal in an **ash monofill** (a landfill containing only ash) would have the least environmental impact, and that fixation followed by monofilling would also have minimal impacts. Ash monofilling without prior treatment would have greater impacts and codisposal with municipal solid waste in landfills without any treatment would be the most detrimental. Volatilization of mercury in an ashfill may also be a problem.

The design and operation of ashfills and landfills is critical in minimizing leaching into groundwater. **Monofills are preferable** because codisposal of ash with municipal solid waste may increase the leachability of the ash by exposing it to acids. The type of lining at the fills is critical: thousands of unlined landfills release millions of gallons of leachate into groundwater supplies daily. **Current state-of-the-art ashfills and landfills must have composite liner systems consisting of multiple (usually two) layers of composite liners (plastic and clay or compacted soil) sandwiched between leachate collection and leak detection systems, with on-site leachate treatment facilities.** Figure 3-12 shows a schematic diagram of a landfill with such systems. Finally, the proper operation of an ashfill is also important, since **daily covering and moistening of the ash are necessary** to prevent dispersion of the ash by the elements.

Ash reuse

The management option of reusing ash is consistent with the increasingly popular idea of recycling waste. This reuse is usually as a construction aggregate, fill, or road paving material.

However, ash reuse as an alternative to disposal raises some serious and controversial issues about ash toxicity and human exposure. Because toxic metals do not

Figure 3-12 *Schematic diagram of a landfill with composite liners and leachate collection systems. (Adapted from: Office of Technology Assessment, 1989, after 52 Federal Register 20226, May 29, 1987)*

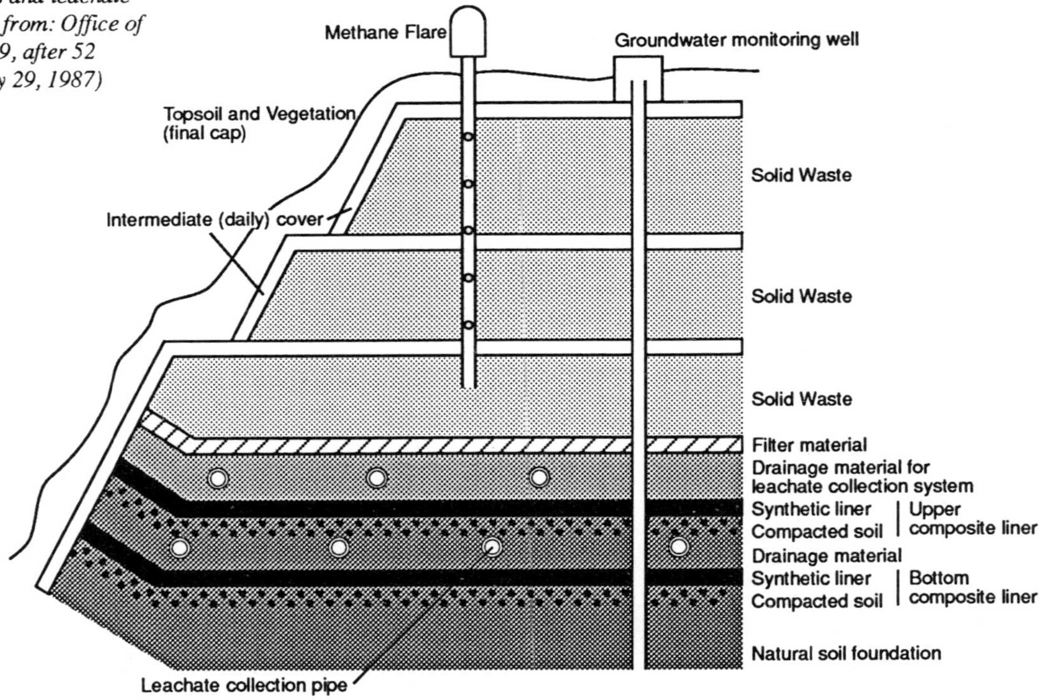

deteriorate, they remain available to re-enter the environment through erosion. Some groups that have studied ash reuse, such as the Environmental Defense Fund, believe that the possibilities of releasing such toxins into the environment inevitably increase by using ash in road building or construction.[29] The use of ash in building materials raises the issue of dust emissions during building maintenance and, particularly, demolition. These environments cannot easily be monitored to control fugitive emissions over the long term.

Some academic ash experts maintain that heavy metals in ash are chemically bound when ash is cemented and that lab tests show that the metals cannot be unbound except under extremely acidic conditions not typical of real-life situations.[30] Thus the metals could not leach into the environment. Essentially, it is not currently certain that these metals will remain immobilized so that they will not be harmful when they do re-enter the environment.

Thus, at this time, ash reuse must not be seen as an escape from more stringent ash disposal regulations. Indeed, ash reuse regulations must be even more thorough in light of the increased human exposure to ash that is not limited to a defined disposal site. Ash reuse may become a viable option in the future as more technologies come on line, as tests are conducted to determine long-term environmental impacts, and as standards for ash treatment and reuse of treated ash are developed and enforced.

Retrofitting Existing Plants to Meet the State of the Art

As state-of-the-art technologies for preventing and reducing emissions from municipal solid waste incinerators improve, and as public pressure continues to mount for their use on both new and existing facilities, emissions regulations are becoming more stringent in Europe and the United States. In turn, these regulations have led to the retrofit of many acid gas air pollution control systems in Europe and will probably lead both to installation of acid gas scrubbers and to upgrading of particulate control equipment in the United States in future years. In some cases, retrofitting to improve combustion and reduce emissions of oxides of nitrogen may also result.[31] Additionally, new federal standards for garbage-burning plants, promulgated by the Environmental Protection Agency in early 1991, include emissions guidelines for existing plants.

Since most older United States resource recovery facilities obtained operating permits when there were fewer and less stringent emission control regulations, typical emissions control equipment for such facilities consists only of a two-stage electrostatic precipitator with no control of acid gases or oxides of nitrogen and no capacity for condensation of flue gases. In terms of the combustion equipment, some older facilities were constructed with traveling grates (which do not agitate the solid waste, resulting in lower combustion efficiency), without proper controls either for admitting the appropriate quantities of air into the primary and secondary combustion areas or for creating optimal turbulence and mixing, without bull noses and arches designed into the furnace for maximizing residence time and mixing, without auxiliary burners or afterburners for achieving and maintaining adequate temperature, without automatic combustion controls, and without a full array of continuous process and emission monitors.

In order to meet increasingly stringent emissions control requirements, some facilities will be faced with adding new and/or upgrading existing equipment or shutting down operations. This choice will be based on case-by-case considerations relating to the specific modifications required at an individual facility: their magnitude and cost, the availability of space for additional equipment, and the cost of and priority given to waste prevention and management alternatives in the community. For example, Dade County (Florida) officials decided to rehabilitate a poorly functioning incinerator, while those in Hempstead (New York) chose to raze a similar plant.

A variety of post-combustion emission control technologies have been successfully retrofitted to existing solid waste incinerators in the United States, Canada, and Europe, allowing them to meet most emission requirements. These include wet scrubbing, spray drying absorption, dry injection scrubbing (with and without activated carbon), and dry injection with humidification (all for control of acid gases); condensers, selective noncatalytic reduction (such as Thermal De-NO_x, and NO_xOUT), and selective catalytic reduction (for control of oxides of nitrogen); multistage electrostatic precipitators and fabric filters (for particulate control); and combinations of the above. Each type of technology has advantages and disadvantages, and site-specific design considerations must be taken into account.

Technical design considerations for evaluating retrofit systems fall into two main categories: (1) levels of emissions control desired or required and (2) space availability (both on the site and within the plant's physical configuration). Additionally, system compatibility is a factor. For example, if a plant has a fabric filter for particulate collection, a wet scrubber could not be placed before it for the acid gas control system.

Emissions Control Levels

The levels of emissions control required for different pollutants for a retrofitted facility will dictate the choice of air pollution control systems to be considered. The more stringent the requirements, the more restricted the choice of technologies will be; more flexibility is possible if less stringent emissions levels are required.

For many existing plants, the largest alteration will involve the addition (or upgrading) of the acid gas/particulate removal system. If the best possible emissions levels are required for acid gases (sulfur dioxide and hydrogen chloride), organic compounds, and heavy metals, a basic retrofit system would consist of either a five-field electrostatic precipitator followed by a wet scrubber or a spray-dry absorption or dry injection scrubber with humidification and a fabric filter. Fabric filter systems have most reliably provided the lowest possible particulate emissions levels (especially of fine particulates), although four- and five-field electrostatic precipitators may also approach these levels. Dry injection scrubber systems without humidification would be eliminated from consideration due to lower removal efficiency.

In addition to the basic scrubber/particulate collector system, other modifications would be needed to achieve the lowest possible emissions of acid gases, organics, and heavy metals. To maximize condensation of these substances, the temperature of the flue gas as it enters the particulate collector must be kept as low as possible: a humidifier or heat exchanger would be required if the existing scrubber did not provide sufficient cooling of flue gases. This increased condensation, in turn, would require special particulate collector design features (such as insulation) to minimize plugging and corrosion, and possibly a reheater for the flue gas to aid in gas dispersal out of the plant's stack. Further possible modifications include partial in-plant recycling of spent scrubber reagent to enhance sulfur dioxide absorption and minimize reagent consumption, and injection of special reagents (such as sodium sulfide or activated carbon) to achieve high mercury and dioxin capture.

Beyond acid gas and particulate control, a retrofitted plant requires some method for controlling emissions of oxides of nitrogen. An ammonia or urea injection system can be added, either in the furnace or after the particulate collector.

Additionally, incorporation of a bed of activated carbon at the end of the pollution control equipment could reduce emissions of dioxins, mercury, oxides of nitrogen, and acid gases even further. Such systems have been demonstrated in pilot plant retrofits in Germany and Austria and, due to heightened concerns about the level of dioxin contamination of cow's milk, some plants in the Netherlands are likely to be retrofitted with several of them. If a plant's combustion is reasonably efficient, use of these various retrofit technologies could result in emissions close to state-of-the-art levels.

Space Availability

The availability of space for added emissions control equipment is an important factor. Depending on the configuration of the existing boiler and pollution control equipment, retrofit of additional devices may be routine, difficult, or impossible. Until recently, plants were usually not designed with space purposely left vacant for possible later inclusion of emission control equipment, although this strategy has been utilized more and more as vendors and owners become aware of changing and increasingly stringent emission regulations.

For incinerators where space is limited between the existing electrostatic precipitator, fan, and stack, the choice of control technology may also be quite limited. Specifically, while wet scrubbers do not require a large "footprint" if no secondary effluent treatment is required, they do typically require 20 to 40 feet of headroom. They are also quite heavy and hence must be located at ground level. Semi-dry systems, on the other hand, generally require more space than wet scrubbers, but their operating weights are much less and they can be installed at the roof level without much difficulty. In those cases where there is little space, if an existing electrostatic precipitator can be upgraded adequately and acid gas control requirements are not stringent, the dry injection process may be the technology of choice. A reagent silo can be located some distance away from crowded areas and the reagent delivered pneumatically to the flue gases before they enter the electrostatic precipitator.

Control technologies for oxides of nitrogen that involve furnace injection are similar to dry injection in that they consist of reagent silos and ductwork and, thus, can fit into small spaces. The same is true for injection of other adsorbents such as activated carbon and sodium sulfide.

Policy Considerations

Despite the availability of such technologies for effective retrofits, and operating experience with them in Europe, the Environmental Protection Agency's new guideline for existing plants do not systematically require their use. Rather, the guidelines establish emissions levels for existing waste-to-energy plants that are less stringent than the standards mandated for new plants, which themselves fall short of the state-of-the-art levels identified by INFORM. (Chapter 6 discusses the new federal regulations in more detail.) Further, states are not required to adopt these guidelines.

The cleanest possible incineration would be obtained by requiring existing facilities to retrofit to state-of-the-art standards. Given both the potential public health impact of incinerator emissions and the cost of a major retrofit, the question of what should trigger a retrofitting requirement — that is, how far short of state-of-the-art standards a plant must be — is a policy issue deserving serious discussion.

Notes

1 Environment Canada, National Incinerator Testing and Evaluation Program, "Air Pollution Control Technology," *NITEP Report*, September, 1986.

2 See, for example, Herbert Needleman, *et al.*, "Deficits in Psychological Classroom Performances of Children with Elevated Dentine Lead Levels," *New England Journal of Medicine*, vol. 300, 1979, p. 689.

3 A. K. Ahmed, and Perera, F., *Respirable Particles*, 1979, pp. 60-61.

4 Environmental Defense Fund, *The Hazards of Ash and Fundamental Objectives of Ash Management*, New York, 1988.

5 Personal communication, Floyd Hasselriis, P.E., to Dr. Maarten de Kadt, INFORM, June, 1990.

6 US Environmental Protection Agency, *Characterizations of Municipal Solid Waste in the United States: 1990 Update*, June, 1990, Figure ES-5.

7 Final Report of the New Jersey Solid Waste Assessment Task Force, August 6, 1990, p. 28.

8 Gershman, Brickner, and Bratton, Inc., *Small-Scale Municipal Solid Waste Energy Recovery Systems*, Van Nostrand Reinhold Company, Inc., 1986, p. 17.

9 The Fredonia Group, "Solid Waste Management," cited in *Integrated Waste Management*, April 3, 1991.

[10] INFORM (Allen Hershkowitz and Eugene Salerni), *Garbage Management in Japan*, New York, 1987.

[11] Edward J. Sommer, *et al.*, "Emissions, Heavy Metals, Boiler Efficiency, and Disposal Capacity for Mass Burn Incineration with a Presorted MSW Fuel," 81st Annual Meeting, Air Pollution Control Association, Dallas, June 19-24, 1988.

[12] Environment Canada, "The National Incinerator Testing and Evaluation Program: Environmental Characterization of Mass Burning Incinerator Technology at Quebec City," Report EPS 3/UP/5, June, 1988.

[13] "Results of the Combustion and Emissions Research Project at the Vicon Incinerator Facility in Pittsfield, Massachusetts," Final Report, vol. I, MRI Project #8949-L(12), June 3, 1987.

[14] W. P. Linak, *et al.*, "Waste Characterization and the Generation of Transient Puffs in a Rotary Kiln Incinerator Simulator," Proceedings of the 13th Annual Research Symposium on Land Disposal, Remedial Action, Incineration, and Treatment of Hazardous Waste, Cincinnati, Ohio; EPA Report #600/9-87-015, July, 1987.

[15] "Air Pollution Control at Resource Recovery Facilities," California Air Resources Board, May 24, 1984.

[16] Jeff Hahn and Sofaer, Donna, "Variability of NO_x Emissions from Modern Mass Fired Resource Recovery Facilities," 81st Annual Meeting, Air Pollution Control Association, Dallas, June 19-24, 1988.

[17] Jack D. Lauber and Drum, Donald A., "Best Controlled Technologies for Regional Biomedical Waste Incineration," presented at the 83rd Annual Meeting of the Air and Waste Management Association, Pittsburgh, PA, June 27, 1990.

[18] Personal communication, Dr. Donald A. Drum, Chairman, Technology/Natural Sciences Division, Butler County Community College, to Dr. Maarten de Kadt, INFORM, June 19, 1990.

[19] "Results of the November 3-6, 1987 Performance Test on the No. 2 RDF and Sludge Incinerator at the WLSSD plant in Duluth, Minnesota," vol. I, Interpoll Laboratories Report #7-2443, April 25, 1988.

[20] Solvie Herstad and Kullendorf, A., "Waste Incineration by Fluidized Bed Technology — Test Results and Experience," Proceedings Municipal Waste Incineration, Environment Canada, NITEP, Montreal, October 1-2, 1987.

[21] Personal communication, David Raring, Raring Corporation, Petaluma, CA, to Marjorie J. Clarke, INFORM, September 1988.

[22] Environment Canada, "The National Incinerator Testing and Evaluation Program: Air Pollution Control Technology (Quebec City)," Report EPS 3/UP/2, 1986.

[23] *Ibid.*

[24] M. D. McDaniel, *et al.*, "Air Emissions Tests at Commerce Refuse-to-Energy Facility — May 26 - June 5, 1987," for County Sanitation Districts of Los Angeles County, Whittier, California, by Energy Systems Associates, July, 1987.

[25] J. Pohl, "Review of Japanese Incinerator Technology," International Workshop on Municipal Waste Incineration, National Incinerator Testing and Evaluation Program, Environment Canada, Montreal, Canada, October 1-2, 1987.

[26] "Recommendations for Policy and Regulations for Residue from MSW Incineration," Toxic Substance Control Commission, State of Michigan, August, 1988.

[27] Environmental Defense Fund, *The Hazards of Ash and Fundamental Objectives of Ash Management,* New York, 1988.

[28] Personal communications, Dr. Donald A. Drum, Butler County Community College, and Floyd Hasselriis, P.E., to Dr. Maarten de Kadt, INFORM, June, 1990.

[29] Environmental Defense Fund, *op. cit.*

[30] Personal communication, Dr. Donald A. Drum, Butler County Community College, to Dr. Maarten de Kadt, INFORM, June, 1990.

[31] Some of the material for this section is derived from James R. Donnelly, "Design Consideration for MSW Incinerator APC Systems Retrofit," Proceedings of the 83rd Air and Waste Management Association Annual Meeting, Pittsburgh, PA, June 24-29, 1990.

CHAPTER 4: ENVIRONMENTAL PERFORMANCE OF 15 INCINERATORS

The 15 waste-to-energy plants INFORM studied vary widely in technologies used and impact on the environment. By comparing their environmental performance, we can draw some conclusions about which technologies are most effective in minimizing the negative environmental impacts of garbage-burning. We can also distinguish between problems that are largely solved and problems for which the solutions remain to be found or implemented. And, in a broader sense, we can identify issues that must be addressed if incineration is to play the most environmentally sound role possible for municipalities that choose to make it a part of their overall waste management strategy. If, after adopting aggressive source reduction and recycling programs, communities decide to burn the remaining waste, they can ensure the safest, cleanest incineration if they require the best possible equipment and techniques.

The INFORM Study

State-of-the-art incineration technology is constantly evolving. INFORM embarked on this study of 15 waste-to-energy plants to examine the wide range of technology and practices actually in use in the United States. We wanted to compare the plants and their environmental impacts with each other and with state-of-the-art technology and emissions levels.

The 15 plants that INFORM selected for the study mirror the diversity of the 128[1] waste-to-energy facilities operating in the United States at the end of 1990: varying features include ownership and management, age, size, fuel, furnace design, and emissions control systems. In order to include plants incorporating examples of most types of state-of-the-art equipment (such as the first plant to use the fabric filter/acid gas scrubber combination, and the only operational plant in the United States with an ammonia injection process for control of oxides of nitrogen), we primarily examined recently built incinerators. Thus, although older incinerators exist (built in the 1960's and 1970's), we excluded them from our study. Newer plants that have come on line since our research began are also excluded. All the facilities described were operating at the time of the study.

Two plants require special mention. The Auburn plant in Maine, the oldest plant in this study, was closed in anticipation of complete reconstruction after research for this study was completed. Nevertheless, it is included in this report because it dramatically illustrates the effects of operating without air pollution control devices. The Lakeland, Florida, plant is unique because it is essentially a coal-burning plant with 10 percent of its fuel derived from refuse (RDF). While we can draw some interesting conclusions about its operations (and several other existing plants burn such a mix), its

Table 4-1 Basic Plant Characteristics

Plant	State	Owner Public	Owner Private	Operator Public	Operator Private	Year Operations Began
Albany	NY	■		■		1982
Auburn*	ME	■		■		1981
Baltimore	MD	■			■	1985
Biddeford/Saco	ME		■		■	1987
Claremont	NH		■		■	1987
Commerce	CA	■		■		1987
Dade County	FL	■			■	1982
Lakeland	FL	■		■		1983
Marion County	OR		■		■	1986
Oswego	NY	■		■		1986
Pascagoula	MS	■			■	1985
Pigeon Point	DE		■		■	1987
Tampa	FL	■			■	1985
Tulsa	OK		■		■	1986
Westchester	NY		■		■	1984

NA, Information not provided by plant.
* *Closed in February 1990.*
† *During reconstruction.*
** *Primary fuel (90%) is high-sulfur pulverized coal.*

performance is not directly comparable to that of plants burning only municipal solid waste or refuse-derived fuel.

The study process consisted of plant visits, follow-up questionnaires, and interviews. INFORM visited the plants over a three-year period during which the garbage-burning industry was rapidly changing. Each visit, lasting one or two days, included a plant tour and interviews with plant managers and municipal officials. The interviews followed a standard outline.

Type of Fuel		Size			
		Design capacity (tons/day)	Actual operations (tons/day)	Energy rating	
MSW	RDF				
	■	600	400	NA	
■		200	185	50,000	lb steam/hour
■		2250	2250	60	mw
	■	607	607	22	mw
■		200	171	4.5	mw
■		330	330	11.44	mw
	■	3000	1700 †	76	mw
	■**	500	370	364	mw
■		550	510	13.1	mw
■		200	190	4	mw
■		150	125	32,000	lb steam/hour
■	■	600	390	18	mw
■		1000	850	22.5	mw
■		1125	925	18.2	mw
■		2250	1800	60	mw

As the study progressed, additional issues surfaced. These were incorporated into a follow-up questionnaire: plant managers were asked to respond to the new questions and to confirm the accuracy of the data obtained in the earlier visits. (The plant profiles in Appendix A closely follow the questionnaire INFORM used.) INFORM also carried out extensive telephone interviews with plant, state, and local officials, and with other people knowledgeable about the operations of each plant.

Finally, just before going to press, INFORM checked with each plant in the study for an update on plant status and operations, allowing significant changes in the plants, as of early 1991, to be discussed. Appendix B describes the study methodology in more depth.

The Study Plants

The basic characteristics of the 15 plants, and their structure and operations, form the framework for understanding their environmental impacts and for assessing which factors are associated with different levels of environmental performance. While the details for each plant are presented separately in the profiles at the end of this study (Appendix A), this section summarizes the information to facilitate plant-to-plant comparisons.

Basic Characteristics

The plants studied vary in location, ownership and operation (public or private), year operations started (1981-1987), type of fuel burned (municipal solid waste or refuse-derived fuel), **design capacity** (the maximum amount of fuel a plant was designed to burn), and **energy rating** (the maximum amount of energy it can produce). Table 4-1 summarizes this information. (Throughout this chapter, the names in Table 4-1 are used to identify the 15 plants. Often, these represent the location of the plant, rather than its formal name. The formal name and the specific location of each plant were listed in Table 1-1 and are indicated in the plant profiles in Appendix A.)

The Plant Design Process

Vital to incinerator design, as discussed in Chapter 3, is the determination of its size through a careful assessment of the amount and type of wastes to be burned. As Table 4-2 indicates, nine of the 15 plants assessed the composition of the community waste stream in the planning process, with six of them indicating they used sampling, the preferred method for obtaining composition data. (However, the information provided to INFORM did not always make it clear what sampling methodologies were used.) In some cases, the sampling method may not have accurately reflected the waste composition of the community to be served by the incinerator. The Commerce plant, for example, based estimates on sampling of the entire Los Angeles area rather than the immediate local industrial community from which it receives its waste, and the Pigeon Point plant used waste composition data from 1978 for a plant that began operations in 1986. Only four of the plants conducted elemental analyses of the waste stream that could reveal the presence of toxic materials or pollutant precursors: Biddeford/Saco, Commerce, Pascagoula, and Pigeon Point.

Effective plant design incorporates the impact of existing and future recycling and source reduction programs into the waste volume and composition studies since recycling and reduction decrease the amount of waste to be burned and alter the kinds of materials in the waste stream (and hence combustion quality). While recycling now exists in most of the communities served by the plants in this study, it was not considered in the planning process for any facility. Incinerator standards proposed by the Environmental Protection Agency in 1989 would have required communities to achieve 25 percent source reduction/recycling/composting before incineration permits were granted. However, this requirement was omitted from the final version of the regulations.

Table 4-2 also presents information about siting the plants. It is notable, in this era of "Not in My Back Yard," or NIMBY, that 13 of the plants are 1 mile or less from the nearest residence (none is more than 2 miles away) and that plant operators at nine of the plants told INFORM they had experienced no community opposition at the planning stages. A possible explanation is the relative lack of awareness of potential environmental impacts at the time of plant construction and the relatively recent establishment of NIMBY as a powerful phenomenon. There has, in fact, often been more community opposition after the plants began operations. (However, this information is anecdotal; INFORM did not systematically seek data on post-construction opposition.) Further, only four were built at completely new locations; the others are either on the site of existing waste disposal or steam-producing facilities or near the plant's energy customer.

The apparent ease of siting most of these 15 plants cannot be expected to continue into the future. The Citizen's Clearinghouse on Hazardous Waste reported that, in the fiscal year 1990, 107 garbage-burning plants, or plans to build them, were blocked, shut down, cancelled, or delayed, primarily by community opposition.[2]

Plant builders are finding it increasingly difficult to obtain operating permits. To take just one example, by the middle of 1991, no permit had been issued for New York City's Brooklyn Navy Yard plant, the first of eight proposed incinerators, although the city's Board of Estimate had approved the project in 1984 and planning began in the late 1970's. Perhaps the pressure of the municipal solid waste crisis will pave the way for additional garbage-burning plants, but it is also possible that public sentiment will lead to increased reliance on source reduction and recycling programs.

Screening Wastes

Every plant in INFORM's study reported some method for screening wastes brought to the facility and identifying prohibited wastes. Except for the four refuse-derived fuel plants, which remove such noncombustible materials as metals, glass, and grit, the plants attempt to screen only for the general types of bulky and potentially harmful materials listed in Table 3-1.

The extent to which the mass burn plants succeed in such screening is limited. At Tulsa, for example, INFORM saw a charred but incompletely burned tire and some tire-sized metal objects come out of the plant with the combined bottom and fly ash. Further, the amounts of wastes actually rejected are small. Only four of the 12 mass burn plants reject 1 percent or more of the incoming wastes, and none rejects more than 3 percent. Considering that 15.5 percent of the United States waste stream by weight consists of metal and glass,[3] much more waste could be rejected simply on the criterion of noncombustibility.

Table 4-2 Plant Planning

Plant	Criteria for Sizing and Design/Study Method		
	Volume	Composition	Elemental analysis
Albany	Weight records	Sampling	NA
Auburn	Waste haulers' estimates	NA	No
Baltimore	Weight records	None	No
Biddeford/Saco	Method not supplied	None	Yes
Claremont	Weight records	Sampling	No
Commerce	Sampling*	Sampling	Yes
Dade County	Weight records	None	No
Lakeland	Weight records	NA	NA
Marion County	Weight records	Sampling	No
Oswego	Method not supplied	Method not supplied	No
Pascagoula	Landfill records	Method not supplied	Yes
Pigeon Point	Estimate, weighing	Sampling (in 1978)	Yes (1978)
Tampa	Weight records	Method not supplied	No
Tulsa	Weight records	None	No
Westchester	Weight records	Method not supplied	No

NA, Information not provided by plant.

* *Estimates based on sampling of entire Los Angeles area; City of Commerce is an industrial area.*

† *Opposition to first proposed site; agreement on current site.*

Siting		Year	Citizen	Community
Nearest residence	Location	Operations Began	Involvement in Planning	Opposition at Plannning Stage
1 block	Existing steam-producing facility	1982	Public hearings	Yes
0.25 mile	Near steam customer	1981	None	Yes
0.5 mile	Site of previous facility	1985	None	No
400 feet	Near steam customers	1987	Municipal planning committee	Yes
0.5 mile	New	1987	Advisory committee	No
900 feet	New	1987	Advisory committee	No
1 mile	Site of existing dump	1982	None	No
2 miles	Existing power plant	1983	Public meetings	No
0.25 mile	New	1986	Advisory committee	Yes
0.5 mile	New	1986	Public meetings	No
2 miles	Near steam customer	1985	Citizen task force	Some†
0.5 mile	Site of disposal area	1987	Workshops, public hearing	No
0.5 mile	Site of previous incinerator	1985	Advisory committee	No
5 blocks	Near steam customer	1986	Open meetings	No
0.5 mile	New	1984	Public meeting	Some

Table 4-3 Identification of Prohibited Wastes

Plant	Method of Identification: Visual inspection	Random sampling	Radioactivity meter	Hand-picked	Staff Responsible for Identification
Mass Burn					
Auburn	■				Loader operators
Baltimore	■				Crane operators
Claremont	■				Loader operators
Commerce	■	■	■		Weigh scale, bulldozer, and crane operators
Marion County	■				Crane operators
Oswego	■				Loader operators
Pascagoula	■				Furnace loaders
Pigeon Point*	■				Floor inspectors
Tampa	■	■			Crane operators, plant manager, local gov't. representative
Tulsa	■	■			Plant workers, local government
Westchester	■				NA
Refuse-Derived Fuel					
Albany	■			■	Tip floor workers
Biddeford/Saco	■				Tip floor workers
Dade County	■				Waste management officials
Lakeland	■			■	Tip floor workers

NA, Information not provided by plant.
—, Not relevant
* *Pigeon Point burns* both *municipal solid waste and refuse-derived fuel in varying proportions.*
† *A recent revision to the City Code also allows the City of Tampa to bill responsible party for clean-up of prohibited waste.*
** *Figure includes refuse-derived fuel rejections: at Biddeford/Saco, 0-1% rejected after RDF process; at Dade, 2% rejected after RDF process.*

Percent Rejected	Penalties Type	Ever applied	Enforcement
3	None	—	—
< 1	Warnings/bans possible		Plant personnel
< 1	Warnings/exclusion possible		NA
1.6	$25,000 fine, loss of license		Local gov't
0.5	Loss of license possible for repeat offenders		State gov't
2-3	Rejection of entire load	■	Plant personnel
0.01	Reprimand	NA	NA
1-2	Offenders referred to state environmental department		State gov't
< 1	Prohibited waste returned†	■	Local gov't
0.5	Offenders must remove load	■	Local gov't
< 1	None	—	—
5	Suspension of permits	NA	Local gov't
18-20**	Exclusion of carter		State gov't
40**	Fines		Local gov't
10	None	—	—

As Table 4-3 demonstrates, with the sole exception of a radioactivity meter at the Commerce plant, these plants rely on visual inspection by the crane or front-end loader operator who is mixing the wastes, or by other pit or tipping floor workers, with occasional random sampling at Commerce, Tampa and Tulsa. (At the time INFORM visited Commerce, every fifth load was dumped on the tipping floor to facilitate inspection.) Once wastes reach the plant, the exclusion of prohibited wastes thus depends largely on the training and continued vigilance of plant personnel. However, this is clearly not always enough. As the control room operator at the Tampa plant told INFORM, "How could [the crane operator]... spot a lead acid battery from here [high above the garbage pit]?"[4]

Penalties for bringing prohibited wastes to the plants are limited, and the extent of enforcement is even more limited. Twelve of the 15 plants have some form of penalty but only two report ever imposing one. Two of the 15 plants can levy fines on violators (Commerce and Dade County); these fines have never been imposed. Six of the plants report that violators can be excluded from the plant in the future, but none reports ever imposing this sanction. Four of the other plants claim to reprimand the violators or require them to remove the load with prohibited wastes; only Oswego, Tampa, and Tulsa report ever imposing these penalties. Three plants (Auburn, Lakeland, and Westchester) have no penalties at all.

Plant Structure

A waste-to-energy plant is a complex system, consisting of several variable components (storage location, loading mechanism, furnace, heat-energy transfer system, emissions control equipment, and ash collection process), the design of which can affect the facility's environmental performance. Table 4-4 summarizes each plant's structural characteristics for easy comparison.

The plants are approximately evenly divided in their use of the tipping floor storage system that improves plant workers' ability to observe prohibited wastes and the pit system. All six of the plants that rejected 1 percent or more of the incoming wastes have tipping floors except Albany; however, Albany is a refuse-derived fuel plant with its pit storing refuse-derived fuel that is prepared off-site. Further, only one plant with a tipping floor rejected less than this amount: Claremont.

Eleven of the plants have a continuous loading system. As discussed in the technology chapter, continuous loading permits more even combustion conditions than batch loading. The four plants with batch loading are Auburn, Biddeford/Saco, Oswego, and Pigeon Point.

While dual-chambered furnaces are designed to increase combustion efficiency and reduce pollutant formation, automatic combustion controls, auxiliary burners, and internal furnace design can improve the performance of single-chambered furnaces. Single-chambered furnaces predominated in the study (10 of 15), as they do throughout the United States. However, the sample does include four dual-chambered furnaces and two plants with rotary kiln furnaces (Pascagoula, which has two chambers as well as a rotary kiln, and Tampa). Eleven of the plants have auxiliary burners, including eight of the 10 with single-chambered furnaces, and 11 have automatic combustion controls. The plants with more than one chamber use controlled rather than excess air, with the exception of Pascagoula.

Emission control equipment also varies. Five plants have scrubbers of the types that represent state-of-the-art technology for acid gas reduction; six (including Lakeland, which is primarily a coal-burning plant) have fabric filters or electrostatic precipitators with four or more fields, considered state-of-the-art technology for particulate control in waste-to-energy facilities; eight have precipitators with less than four fields; and only two (Commerce and Pigeon Point) have systems for controlling emissions of oxides of nitrogen. The recently closed Auburn plant operated with no functioning add-on pollution control equipment at all. It is important to remember, however, that how such emissions control devices are operated affects their performance.

Monitoring and Maintenance

The existence of automatic combustion controls does not obviate the need for monitoring the burning process by skilled control room operators. Such monitoring requires the existence of a control room with sufficient information presented in a convenient form to permit operator intervention. Operators trained in the interpretation of the information presented are also required.

Control rooms vary widely, and most are far from state of the art which would allow the control room operator to monitor all parameters simultaneously. The Tampa plant is noteworthy because the control room operator and the crane operator are located in the same place and can thus be in constant direct communication with each other. In some control rooms, it is not possible to monitor all parameters continuously; in others, readings of all parameters are not simultaneously visible. At Biddeford/Saco, for example, operators must depress switches to change screens on the monitors to report furnace temperature. In the Commerce plant, INFORM saw control room operators leaving the control room to observe furnace conditions directly. The control room at Oswego is not staffed, although employees periodically check the monitoring equipment, and the recently closed Auburn plant had no control room at all (although temperature and steam pressure readings were available on panels in front of the combustion units).

Table 4-5 summarizes continuous monitoring information about the study plants. Of the nine operating and emissions monitors identified by INFORM as necessary for state-of-the-art monitoring, the table lists only seven, since the information provided by the plants did not consistently distinguish between furnace and flue gas temperature or consistently name steam flow. Thus, temperature is listed generally, and steam flow is not included.

None of the plants had continuous monitors for all seven operating and emissions factors recommended and surveyed by INFORM at the time the research was carried out, although Commerce, Pigeon Point, and Westchester had six each. (Commerce has subsequently added an opacity monitor and now has all seven.) All the plants monitor temperature and most also measure steam pressure and oxygen. Along with carbon

Table 4-4 Plant Structure

Plant	Storage: Pit	Storage: Floor	Percent of Incoming Waste Rejected	Loading: Continuous	Loading: Batch	Furnace: Type	Furnace: Air: Controlled
Albany	■		5	■		Single-chamber	
Auburn		■	3		■	Dual-chamber	■
Baltimore	■		< 1	■		Single-chamber	
Biddeford/Saco		■	18-20*		■	Single-chamber	
Claremont		■	< 1	■		Single-chamber	
Commerce	■†		1.6	■		Single-chamber	
Dade County**		■	40*	■		Single-chamber	
Lakeland		■	10	■		Single-chamber	
Marion County	■		0.5	■		Single-chamber	
Oswego		■	2-3		■	Dual-chamber	■
Pascagoula	■		0.01	■		Dual-chamber, rotary kiln	
Pigeon Point		■	1-2		■	Dual-chamber	■‡
Tampa	■		< 1	■		Rotary kiln	
Tulsa	■		0.5	■		Single-chamber	
Westchester	■		< 1	■		Single-chamber	

* *Before refuse-derived fuel process: 0-1% rejected after RDF process at Biddeford/Saco; 2% at Dade.*

† *Every fifth load is dumped onto a tipping floor instead of into the pit.*

** *Since the research for this study was conducted, Dade County has added storage pits; the plant has an auxiliary burner that was not operating at the time of the research.*

‡ *Flue gas recirculation.*

			Emissions Control Equipment			
Excess	Auxilliary burners	Automatic combustion controls	Electrostatic precipitator (no. of fields)	Scrubber	Fabric filter	NO_x control
■	■	■	3			
■	■	■	4			
■	■			■	■	
■	■	■		■	■	
■	■	■		■	■	Thermal De-NO_x
■	■	■	3			
■			5	■		
■	■	■		■	■	
	■	■	2			
■	■		2			
	■	■	3			Flue gas recirculation
■		■	2			
■		■	3			
■	■	■	3			

Table 4-5 Continuous Monitoring

Plant	Temperature	Steam pressure	Oxygen	Carbon monoxide	Sulfur oxides	Oxides of nitrogen	Opacity
	Parameters Monitored						
Albany	■	■					■
Auburn	■	■					
Baltimore	■	■	■				■
Biddeford/Saco*	■		■	■			■
Claremont	■	■	■	■			■
Commerce**	■	■	■	■	■	■	
Dade County**	■						■
Lakeland	■	■	■		■		■
Marion County**	■	■	■				■
Oswego	■	■	■	■			
Pascagoula	■	■					■
Pigeon Point	■	■		■	■	■	■
Tampa	■	■	■				■
Tulsa	■	■	■	■			
Westchester	■	■	■	■		■	■

NA, Information not provided by plant.

* *Biddeford/Saco also monitors hydrogen chloride, not yet a sufficiently widely accepted technique to be considered a state-of-the-art requirement. Since research for the study was completed, it has added sulfur dioxide and carbon dioxide monitors.*

† *Telemetering link to State Department of Environmental Protection has been established since research for this study was completed.*

‡ *Readings telemetered to remote location.*

** *Since research for this study was completed, Commerce has added an opacity monitor; Dade has added monitors for oxygen, carbon monoxide, and carbon dioxide; and Marion County has added monitors for sulfur dioxide and oxides of nitrogen.*

Monitors Connected to Alarms	Parameters Recorded	Time Records Kept	Frequency of Reporting	Monitor Calibration Frequency
NA	NA	NA	NA	NA
None	None	Not kept	Not required	No schedule
All	All	2 years	Quarterly	Continuous
All	All‡	7 years	No schedule†	Every 6 months
All	All	2 years	Quarterly	Continuous
None	All	NA	Monthly	Daily
None	Opacity	24 hours	Quarterly	Every 8 hours
Opacity, sulfur dioxide	All	5 years (off-site)	Quarterly	Daily
All except steam	All	3 years	Monthly	Daily
All	All‡	NA	NA	Monthly
None	Temperature, opacity	7 years	On request	Every 2-3 months
All	All	NA	NA	NA
All	All	2 years	Quarterly	Quarterly
All	All	5 years	Not required	Annually
Opacity, SO_2, CO	All	3-5 years	Not reported	Daily

Table 4-6 Maintenance Schedules

Plant	Furnace	Boiler	Stoker and Grate
Albany	NA	NA	NA
Auburn	Monthly	Weekly	Quarterly
Baltimore	Quarterly	Quarterly	NA
Biddeford/Saco	NA	Annually	Periodic cleaning, annual maintenance
Claremont	Quarterly	Quarterly	Quarterly
Commerce	NA	NA	NA
Dade County	As warranted by inspection	As warranted by inspection	Every 6 months
Lakeland	Twice a year	Twice a year	Twice a year
Marion County	Twice a year	Twice a year	Twice a year
Oswego	Every 4 weeks	Every 2 weeks	Every 4 weeks
Pascagoula	Every 3 weeks	Every 3 weeks	Every 3 weeks
Pigeon Point	Twice a year	Twice a year	Twice a year
Tampa	Every 4 months	Every 4 months	Every 4 months
Tulsa	Twice a year	Twice a year	Twice a year
Westchester	Every 3-4 months	Every 3-4 months	Every 3-4 months

NA, Information not provided by plant.

monoxide (monitored by seven plants), temperature and oxygen measurements indicate combustion efficiency and can thus provide a guide to the release of products of incomplete combustion. Eleven of the plants monitor opacity, a crude indicator of particulate levels (12, including Commerce's recently added monitor). Only Commerce, Lakeland, and Pigeon Point monitored sulfur oxides (Biddeford/Saco and Marion County have recently added such monitors), and only Commerce, Pigeon Point, and Westchester monitored oxides of nitrogen (Marion County has subsequently added a monitor). Commerce and Pigeon Point are the only plants with any devices for control of emissions of oxides of nitrogen.

Connecting the monitors to alarms warns operators instantly when a parameter exceeds a predetermined level: ten of the plants have alarms for at least some of the parameters.

Air Pollution Control Equipment	Turbine	Other
NA	NA	NA
NA	NA	Monthly
Quarterly	NA	NA
Weekly inspection, continuous maintenance	Annually	NA
Quarterly	Every 3-5 years	NA
NA	NA	NA
Annually	Every 2 years	Cranes: weekly Processing equipment: monthly
Twice a year	Twice a year	Twice a year
Twice a year	Every 3-5 years	Twice a year, or as necessary
Every 4 weeks	Annually	Normal prevettive maintenance
Every 3 weeks	Every 3 weeks	Every 3 weeks
Twice a year	Twice a year	Twice a year
Every 4 months	Every 5 years	As needed
Twice a year	NA	Twice a year, or as necessary
Every 3-4 months	Every 5 years	NA

Records kept as a result of monitoring permit a review of a plant's operating history; some jurisdictions require periodic reporting of such records. As Table 4-5 indicates, nine plants report that they keep records for two years or more, and seven report the information to an outside agency on a regularly scheduled basis. Only two plants, Oswego and Biddeford/Saco, telemeter the information directly to a remote location (the Oswego Department of Public Works, which owns and operates the plant, and the Maine Department of Environmental Protection, respectively).

Regular maintenance is necessary to maintain operating efficiency and reduced emissions levels, although it does not obviate the need for effective equipment and operating practices. The plants in this study report widely varying regular maintenance schedules for key sections of the facilities: from every 3 weeks to annually for air pollution control equipment, for example (see Table 4-6).

Air Impacts

INFORM examined air emissions of the six pollutants for which state-of-the-art emissions levels were defined in Chapter 3 (particulates, carbon monoxide, dioxins and furans, hydrogen chloride, sulfur oxides, and oxides of nitrogen), and of lead and mercury. In this section, we compare the performance of each of the 15 study plants with that of the other plants and with state-of-the-art levels where applicable, and we attempt to identify the factors associated with high and low emissions levels. (A summary of the emissions of the six key pollutants at all 15 plants studied can be found in Table 4-16 at the end of this section, and in the text accompanying it.)

The case study research method INFORM used was designed to allow comparisons across a wide range of facility types and operating equipment and conditions. Therefore, although INFORM did not attempt to evaluate causal relationships, the patterns that appear can shed light on key issues affecting the cleanest possible operation of waste-to-energy incinerators.

Making these comparisons is not always easy. The technology chapter identifies three stages for minimizing air emissions: first, minimizing the amount of noncombustibles, toxic materials, and pollutant precursors that enter the furnace; second, designing and operating the furnace to enhance combustion efficiency and reduce pollutant formation; and, third and last, designing and operating emissions control equipment to prevent the release of pollutants that have been created. For the most part, we have only been able to relate the emissions levels with emissions control devices because there exists such variation in the combinations of types of equipment in use at waste-to-energy plants. Even with a sample of 15 plants, clear patterns are not discernible. The many different structural features, operational practices, and human factors interact in complex ways, and we can often only speculate about how they contribute to observed emissions levels.

Further, because of inconsistent monitoring, systematic across-the-board comparison of the emissions from all 15 plants was impossible. Stack measurement patterns at these plants were inconsistent and in most cases incomplete. Additionally, the plants did not use standard measurement techniques or units of measurement. As a result, comparison is hampered. In fact, only three facilities (Biddeford/Saco, Commerce, and Marion County) measured all six of the key pollutants for which INFORM established state-of-the-art levels and reported their results in a format that INFORM could compare with other results.

The data used here were obtained from plant managers, the state agency that regulates the plant, or state reports, based on one or several discrete tests. They provide a snapshot of an incinerator's performance at a specific time, but may or may not be representative of the plant's "typical" emissions levels, and do not provide a picture of an incinerator's operation over time. For example, the Biddeford/Saco plant, which achieved state-of-the-art emissions levels for four of the six key pollutants based on one-time stack tests, has experienced fires and blow-outs of its fabric filter system that spewed ash over the surrounding community.

Additionally, testing methodology usually involves several observations during one day or a series of observations over several days. In some cases, the variances between observations are minor; in others, great. Test reports may present these multiple observations as discrete numbers or as averages. Where discrete numbers were provided, INFORM has averaged them to facilitate comparisons in the tables and discussion that follow. Information on the number, dates, and results of individual tests is included in the profile of each plant in Appendix A.

Finally, testing procedures and formats for presenting emissions data are not all standardized; thus, plants provide information using a variety of units of measurement and a variety of test conditions. Where possible, INFORM has standardized the data provided by the plants to permit comparisons among the facilities. Appendix B explains the adjustment procedures used. The figures used in the tables that follow can be assumed to be comparable for the purpose of observing the trends and patterns discussed in this report.

In the tables that follow, the plants are listed in order of environmental performance, generally with the plant with the lowest (best) emissions level listed first. If a plant did not measure a given pollutant, or if the data it provided could not be standardized into a format comparable with that of other plants, this is indicated in the tables.

Emissions Testing

The number of pollutants tested for ranges from a high of 23 for Commerce, the newest plant in the sample, to just one for two of the oldest plants in the study, Dade County and the recently closed Auburn plant. Six of the plants tested for all six of the pollutants for which INFORM established state-of-the-art levels (but only three did so in a format permitting comparison), and four more tested for five of them. Some of the other pollutants plants measured include lead (10 plants), mercury (nine plants), beryllium (eight plants), and arsenic, cadmium, and nickel (five plants for each).

Table 4-7 gives a complete listing of what emissions tests were carried out at the study plants. All of the data used in this study come from stack tests, rather than continuous emissions monitors; the plant profiles in Appendix A provide additional information about the timing of the tests and the number of samples involved, and the tables that follow on emissions of individual pollutants indicate which results were reported in comparable formats. (It should be noted that, to facilitate comparison of emissions test data from different plants, all the data reported in the following tables are based on tests conducted around 1987-1988; in some cases, subsequent emissions tests have taken place.)

Particulates

All 15 plants in the study measured emissions of particulates, the minute particles that are released into the flue gas when noncombustible materials are incinerated and when combustible materials are incompletely burned. Particulates also condense out of the gas leaving the furnace from wastes that are vaporized during burning. Seven of the plants exhibited levels lower than the state-of-the-art emissions level of 0.010 grains per dry standard cubic foot (gr/dscf) identified by INFORM; seven others (the exception

Table 4-7 Emissions Testing

Plant	Total Pollutants Tested	Particulates*	Carbon Monoxide*	Dioxins / Furans*	Hydrogen Chloride*	Sulfur Oxides*	Oxides of Nitrogen*	Arsenic	Beryllium
Commerce	23	■	■	■	■	■	■	■	■
Oswego	16	■	■	■	■	■	■	■	■
Albany	15	■		■	■	■	■	■	■
Westchester	15	■		■	■	■	■	■	■
Pigeon Point	13	■	■	■	■	■	■	■	■
Tulsa	13	■	■	■	■	■	■		■
Marion County	12	■	■	■	■	■	■		■
Biddeford/Saco	10	■	■	■	■	■	■		
Tampa	9	■	■			■	■		■
Claremont	6	■	■	■	■	■	■		
Pascagoula	6	■	■		■	■	■		
Baltimore	6	■	■			■	■		
Lakeland	3	■				■	■		
Auburn	1	■							
Dade County	1	■							

* *State-of-the-art emissions level identified in this study.*

was the now-closed Auburn plant) had levels ranging up to 2.5 times the state-of-the-art level. Compared with emissions for several other pollutants that exceed state-of-the-art levels by orders of magnitude, particulate emissions were relatively well controlled.

Table 4-8 compares particulate emissions levels with plant pollution control devices and the age of the plant. Clearly, fabric filters are effective in reducing particulate emissions: all four of the plants with these devices (Commerce, Marion County, Claremont, and Biddeford/Saco) achieved state-of-the-art emissions levels,

Cadmium	Chromium	Lead	Manganese	Mercury	Nickel	Vanadium	Zinc	Other Pollutants
■	■	■		■	■		■	Hydrocarbons, hydrogen fluoride, total chlorinated hydrocarbons, chlorobenzene, chlorophenol, antimony, copper, selenium, thallium
■	■	■	■	■	■	■	■	
■	■	■	■	■	■	■	■	
■	■	■	■	■	■	■	■	
■	■	■		■	■			
		■		■				Fluorides, hydrogen fluoride, sufuric acid, volatile organic compounds
		■		■				Fluorides, hydrogen fluoride, volatile organic compounds
		■		■				Chlorinated hydrocarbons, volatile organic compounds
		■		■				Fluorides, volatile organic compounds
		■						
								Chlorides, fluorides

whereas only three of the eight plants without them did so. Four-field electrostatic precipitators also appear to be effective, and three-field precipitators may help but are not always enough (only two plants with three-field precipitators, Pigeon Point and Tulsa, achieved the state-of-the-art level). The Lakeland plant might appear to be an anomaly since it has a five-field precipitator but has emissions almost double the state-of-the-art level for garbage-burning plants. However, it must be remembered that Lakeland's fuel is only 10 percent refuse-derived fuel; 90 percent is coal, which produces high levels of particulates when burned.

Table 4-8 Particulate Emissions

Plant	Achieve State-of-the-Art Level (0.010 gr/dscf)	Particulate Emissions (grains per dry standard cubic foot)*
Baltimore	■	0.0024
Pigeon Point	■	0.0041
Commerce	■	0.0043
Marion County	■	0.007
Tulsa	■	0.0072
Claremont	■	0.0077
Biddeford/Saco	■	0.008
Tampa		0.012
Oswego		0.013
Westchester		0.016
Pascagoula		0.018
Lakeland		0.019
Albany		0.020
Dade County		0.0258
Auburn†		0.08

* *Information on test dates and number of samples involved is included in plant profiles in Appendix A for all pollutants.*
† *Closed in 1990.*

That particulate control equipment of any type reduces emissions levels is dramatically demonstrated by the performance of the now-closed Auburn plant. With no operating add-on emissions control equipment at all, its particulate levels were eight times the state-of-the-art level and more than three times the level of the plant with the next highest level. Since Auburn has a dual-chambered furnace, it is likely that uncontrolled particulate emissions at plants with single-chambered furnaces would be even higher.

The age of the plant also seems related to particulate emissions to some extent, since all plants except Baltimore that met state-of-the-art levels began operations in 1986 or 1987, and the oldest plants had the highest emissions levels. However, it is not possible

Emission Control Devices			
Electrostatic precipitator (no. of fields)	Fabric filter	Scrubber	Year Operations Began
4			1985
3			1987
	■	■	1987
	■	■	1986
3			1986
	■	■	1987
	■	■	1987
2			1985
2			1986
3			1984
2			1985
5		■	1983
3			1982
3			1982
			1981

to evaluate how much of this effect is related to age and how much is related to the nature and operations of the emissions control equipment.

Beyond these points, questions remain. Why did the Pigeon Point and Tulsa plants, with three-field precipitators, achieve state-of-the-art levels when other plants with similar equipment did not? Possibly this could relate to design or operational features of Pigeon Point's and Tulsa's relatively newer precipitators; the plants began operations in 1987 and 1986, respectively, compared to 1984 and earlier for Westchester, Albany, and Dade County. Still another factor that could play a role is the size of the collection area of the precipitators.

Finally, the comparatively low variation in particulate emissions levels compared to variations for other pollutants, and the fact that those that exceeded state-of-the-art levels did so by less than the exceedence of other pollutants, support one of the findings of this study. That is, the longer engineers and operators have had experience reducing emissions of a particular pollutant, the better the technology and operations and the greater the success. Particulates are a by-product of many forms of combustion, and planners were able to design waste-to-energy plants to reduce their levels from experience with other types of fuel-burning plants.

Heavy Metals

INFORM did not identify state-of-the-art levels for heavy metals, but we did request information on emissions of lead and mercury. (Provisions of the 1990 Amendments to the federal Clean Air Act require the Environmental Protection Agency to develop standards for mercury, lead, and cadmium emissions from new and existing incinerators by late 1991.)

Five plants measured neither lead nor mercury emissions (Auburn, Baltimore, Claremont, Dade County, and Lakeland). Nine plants reported measuring them, as shown in Tables 4-9 and 4-10, respectively. (Pascagoula, which did not measure mercury, did report lead emissions, but in units of pounds per hour that INFORM was unable to recalculate into a format comparable with the other data.) Heavy metals are present in the flue gas if they are present in the wastes entering the furnace; they tend to condense onto particulate matter as the flue gas cools after leaving the furnace.

For lead, Commerce had the lowest emissions level (0.0042 milligrams per normal cubic meter), and Marion County, Pigeon Point, and Biddeford/Saco ranked second, third, and fourth, respectively. Table 4-9 indicates that the scrubbers and fabric filters play a role in reducing lead emissions, since three of the four plants with the lowest levels of lead emissions have this combination of equipment. This is not surprising, since fabric filters were seen to be effective in reducing particulate emissions, and scrubbers enhance condensation by lowering temperature.

At Pigeon Point, which does not have these devices, the low levels are probably partly related to its three-field precipitator which, as discussed above, effectively captured particulates, and to its dual-chambered furnace. (Tulsa, which also has a three-field precipitator and also achieved the state-of-the-art emissions level for particulates, but has a single-chambered furnace, did not have comparably low lead levels.) Its dual-chambered furnace may reduce particulate entrainment, and the lower primary chamber temperature may reduce heavy metal volatilization. Additionally, the wastes burned in Pigeon Point may contain lower levels of heavy metals since it burns refuse-derived fuel (from which some metals have been removed) as well as municipal solid wastes, and its flue gas recirculation system may reduce the amount of metals via repeated opportunities for condensation.

Turning to mercury, none of the plants in this study use recently developed techniques for controlling mercury emissions such as the addition of carbon to the flue gas. However, INFORM found some correspondence between the quality of mercury emissions performance and particulate control. Five of the six plants with the lowest mercury levels had state-of-the-art particulate levels. Three of these use fabric filters and scrubbers (Biddeford/Saco, Marion County, and Commerce), and two (Pigeon Point and Tulsa) have three-field precipitators.

The scrubbers, which tend to reduce temperatures, play a particularly significant role in mercury emissions reduction since mercury's comparatively low boiling point (674°F) means that lower temperatures are needed for it to condense out of the gas phase and thus be captured by the particulate control devices. (Different compounds of mercury have different condensing ranges; mercury chloride, for example, is easier to condense.)

Both lead and mercury emissions levels exhibited dramatically wide variations, especially compared to particulate levels. While the highest particulate emissions were ten times the lowest (leaving out the closed Auburn plant), the highest lead emissions level was more than 300 times the lowest and the highest mercury level was almost 50 times the lowest.

Since one key to heavy metal control is preventing these metals from being burned, some of the variation may be due to differences in the amounts of metals in the refuse burned, whether because the waste stream varies in different locations, or because the thoroughness of waste presorting and refuse-derived fuel programs (removal of lead acid batteries, for example) varies, or both. In this regard, it may be relevant that the plant with the lowest mercury emissions, Biddeford/Saco, is a refuse-derived fuel plant that may remove much of the mercury-containing metals from the original waste stream, and that Pigeon Point, with the second lowest mercury emissions, burns a mixture of refuse-derived fuel and solid waste. By contrast, Commerce, the newest plant in the study, a plant with the latest pollution-control equipment, had relatively high mercury emissions levels, possibly due to higher mercury levels in the mostly commercial wastes it burns or to its not being operated properly.

Additionally, compared to particulates, heavy metal emissions are a recent concern relating specifically to garbage incineration rather than combustion in general. Effective heavy metal emissions control technology, while evolving, has not yet been fully developed or implemented in municipal solid waste incinerators.

Carbon Monoxide

Five study plants did not measure carbon monoxide emissions, one of the products of incomplete combustion (Albany, Auburn, Dade County, Lakeland, and Westchester), and three of the 10 plants that did reported their levels of emissions of carbon monoxide in a format that does not permit comparison. Table 4-11 presents the levels of the seven plants reporting comparable data on carbon monoxide emissions, along with information on several plant design factors known to contribute to efficient combustion.

While Pigeon Point had by far the lowest carbon monoxide emissions, all but one of the seven plants achieved the state-of-the-art emissions level of 50 parts per million. The one plant not meeting this criterion, Biddeford/Saco, had emissions that were less than twice the state-of-the-art level. Just as particulates have long been known as a by-product of many forms of combustion, low carbon monoxide levels have long been

Table 4-9 Lead Emissions

Plant	Lead Emissions (mg/normal cubic meter)	Year Operations Began	Achieve State of the Art for Particulates
Commerce	0.0042	1987	■
Marion County	0.0250	1986	■
Pigeon Point	0.0294	1986	■
Biddeford/Saco	0.077	1987	■
Westchester	0.15	1984	
Tulsa	0.415	1986	■
Tampa	0.776	1985	
Oswego	0.848	1986	
Albany	1.28	1982	

Reasons for lack of lead emissions data: Pascacagoula, emissions measured, but measurement units not convertible to standardized format; Auburn, Baltimore, Claremont, Dade County, Lakeland, not measured by plants.

Table 4-10 Mercury Emissions

Plant	Mercury Emissions (mg/normal cubic meter)	Year Operations Began	Achieve State of the Art for Particulates
Biddeford/Saco	0.0448	1987	■
Pigeon Point	0.0691	1986	■
Marion County	0.28	1986	■
Tulsa	0.419	1986	■
Albany	0.577	1982	
Commerce	0.58	1987	■
Oswego	0.698	1986	
Tampa	0.931	1985	
Westchester	1.92	1984	

Reasons for lack of mercury emissions data: Auburn, Baltimore, Claremont, Dade County, Lakeland, Pascagoula, not measured by plants.

Emission Control Devices		
Electrostatic precipitator (no. of fields)	Fabric filter	Scrubber
	■	■
	■	■
3		
	■	■
3		
3		
2		
2		
3		

Emission Control Devices		
Electrostatic precipitator (no. of fields)	Fabric filter	Scrubber
	■	■
3		
	■	■
3		
3		
	■	■
2		
2		
3		

Table 4-11 Carbon Monoxide Emissions

Plant	Achieve State-of-the-Art Level (50 ppm)	CO Emissions (ppm)	Year Operations Began
Pigeon Point	■	6.7	1987
Commerce	■	16	1987
Marion County	■	19	1986
Oswego	■	<20	1986
Baltimore	■	20.4	1985
Claremont	■	48.7	1987
Biddeford/Saco		83	1987

Reasons for lack of carbon monoxide emissions data: Pascagoula, Tampa, Tulsa, emissions measured, but measurement units not convertible to standardized format; Albany, Auburn, Dade County, Lakeland, Westchester, not measured by plant.

considered a measure of efficient burning. The technology for achieving good combustion is well known, and control techniques can be incorporated into plant design.

Table 4-11 indicates that all but three of the plants that measured carbon monoxide levels have continuous loading, all have auxiliary burners, and all but one have automatic combustion controls. It is highly likely that these design features play a significant role in enhancing complete combustion, but the available data do not permit comparison with plants that lack these features. Similarly, since only two of the plants for which carbon monoxide levels are reported have a dual-chambered furnace, it is not possible from this set of data to evaluate the impact of dual-chambered furnaces. The data also do not allow an evaluation of other factors known from previous controlled studies to enhance combustion: waste presorting, effective waste mixing, and adequate turbulence and residence times.

Dioxins and Furans

Like carbon monoxide, dioxins and furans are also products of incomplete combustion; they form when materials containing precursors, such as certain chlorine-containing plastics, are incompletely burned. However, unlike carbon monoxide control, control of dioxin/furan emissions requires more than simply enhancing combustion efficiency, because dioxins and furans can reform from the precursors after gases leave the furnace. They can then be condensed and captured in properly designed and operated air pollution control devices, and are thus transferred into the fly ash.

Plant Design			
Loading	Number of furnace chambers	Automatic combustion controls	Auxiliary burners
Batch	2	■	■
Continuous	1	■	■
Continuous	1	■	■
Batch	2	■	■
Continuous	1	■	■
Continuous	1	■	■
Batch	1		■

Six of the 15 plants studied did not measure dioxin/furan emissions at all (Auburn, Baltimore, Dade County, Lakeland, Pascagoula, and Tampa), and two (Claremont and Pigeon Point) reported them in measurement units that were not convertible to a standard format. As Table 4-12 shows, only two of the seven plants reporting dioxin/furan emissions in comparable form achieved the state-of-the-art level of 0.10 nanograms per dry normal cubic meter: Commerce and Biddeford/Saco. (There are many different dioxins and furans, with varying toxicity levels. Through calculations, measured dioxin and furan emissions levels are converted into **Eadon toxic equivalents**, giving a comparative indication of dioxin and furan toxicity levels. The state-of-the-art level is stated in terms of these toxic equivalents.)

Both Commerce and Biddeford/Saco, as well as the plant with the next lowest emissions, Marion County, have scrubbers and fabric filters. Further, plants without scrubbers and fabric filters had dramatically higher dioxin/furan levels. While the plant with the third lowest emissions level (and a scrubber and fabric filter) had emissions less than twice the state-of-the-art level, the plant with the fourth lowest emissions (and no scrubber or fabric filter) had a level 17 times that of the state-of-the-art.

Factors that enhance combustion play a role in reducing the primary formation of dioxins and furans, but Table 4-12 shows that such factors were not sufficient to reduce actual emissions to state-of-the-art levels. With the exception of Tulsa, all the plants reporting dioxin/furan levels have furnace temperatures of 1800°F and above and auxiliary burners, and all but one have automatic combustion controls.

Table 4-12 Dioxin/Furan Emissions

Plant	Achieve State-of-the-Art Level (0.10 ng/dNm^3)	Emissions of Dioxin/ Furan Equivalents (ng/dry normal cubic meter)	Year Operations Began
Commerce	■	0.027	1987
Biddeford/Saco	■	0.0712	1987
Marion County		0.155	1986
Tulsa		1.735	1986
Westchester		4.16	1984
Oswego		11.23	1986
Albany		18.8	1982

Reasons for lack of dioxin/furan emissions data: Claremont, Pigeon Point, emissions measured, but units not convertible to standardized format; Auburn, Baltimore, Dade County, Lakeland, Pascagoula, Tampa, not measured by plant.

Thus, scrubbers with efficient particulate collection are critical to reducing emissions of dioxins and furans, although they are not sufficient for achieving state-of-the-art levels. Without scrubbers, emissions as high as 188 times the state-of-the-art level were reported. Scrubbers generally reduce the temperature of the gases leaving the boiler, and low temperatures aid in condensation and capture of dioxins and furans.

As for other pollutants of more recent concern, the technology for effective dioxin/furan emissions control is not yet in widespread use. Source reduction and presorting to remove metal catalysts (such as copper) and precursor-containing materials (such as certain plastics and papers) may also play an important role in reducing these emissions in the future.

Acid Gases

Emissions of hydrogen chloride, an acid gas formed when chlorine-containing wastes (such as some papers and plastics, yard and food wastes, and salt) are burned, were measured in a comparable format at nine plants. Five plants (Auburn, Baltimore, Dade County, Lakeland and Tampa) did not measure hydrogen chloride emissions, and Pascagoula measured them, but reported them using nonstandard units.

Three of the nine achieved the state-of-the-art emissions level of 25 parts per million (Commerce, Biddeford/Saco, and Marion County), and one (Claremont) had emissions less than three times this level. All four of these plants, as Table 4-13 shows, have scrubbers, while the plant with the next lowest emissions (and no scrubber) had

Emission Control Equipment					
Electrostatic precipitator (no. of fields)	Fabric filter	Scrubber	Furnace Temperature (°F)	Auxilliary Burners	Automatic Combustion Controls
	■	■	1800°	■	■
	■	■	1800-2000°	■	
	■	■	1800° (min)	■	■
3			1745° (min)		■
3			2000° (min)	■	■
2			1800°	■	■
3			2500°	■	■

emissions more than 16 times the state-of-the-art level. Dry scrubbers operate by condensing acid gases and/or neutralizing them with alkaline materials.

Sulfur dioxide is another acid gas; it is formed when sulfur-containing materials (such as tires, gypsum board, and shingles) are burned. All but two plants (Auburn and Dade County) measure sulfur dioxide emissions, but one (Pascagoula) uses measurement units that could not be converted to the standard format. As Table 4-14 shows, scrubbers were also effective in minimizing these emissions. The three plants with the lowest emissions, including the two that achieved the 30 parts per million state-of-the-art level (Biddeford/Saco and Commerce), have scrubbers; the sulfur dioxide emissions of that third plant were less than 1.5 times the state-of-the-art level.

However, scrubbers alone were not sufficient: the Claremont plant, with a dry injection scrubber, had sulfur dioxide emissions almost three times the state-of-the-art level. It is possible that the use of a dry injection scrubber (which has been associated with higher sulfur emissions), as well as some aspects of the operation of this incinerator (exit temperature or scrubber lime/acid ratios, for example) account for this. And, while the Lakeland plant, with sulfur dioxide emissions almost three times the state-of-the-art level, has a wet scrubber, 90 percent of its fuel is high-sulfur coal, not solid waste, so it is not possible to directly compare its sulfur dioxide emissions levels to those of the other plants.

Both hydrogen chloride and sulfur dioxide emissions levels showed a wide range

Table 4-13 Hydrogen Chloride Emissions

Plant	Achieve State-of-the-Art Level (25 ppm)	Hydrogen Chloride Emissions (ppm)	Year Operations Began
Commerce	■	8.9	1987
Biddeford/Saco	■	10	1987
Marion County	■	11.9	1986
Claremont		70.3	1987
Tulsa		412	1986
Albany		464	1982
Pigeon Point		541	1987
Oswego		552	1986
Westchester		646	1984

Reasons for lack of hydrogen chloride emissions data: Pascagoula, measured, but measurement units not convertible to standardized format; Auburn, Baltimore, Dade County, Lakeland, Tampa, not measured by plant.

Table 4-14 Sulfur Dioxide Emissions

Plant	Achieve State-of-the-Art Level (30 ppm)	Sulfur Dioxide Emissions (ppm)	Year Operations Began
Commerce	■	1.3*	1987
Biddeford/Saco	■	5	1987
Marion County		42	1986
Tampa		78.8	1985
Tulsa		96.0	1986
Baltimore		119	1985
Westchester		140	1984
Claremont		145	1987
Pigeon Point		163.89	1987
Lakeland†		179.2	1983
Albany		224	1982
Oswego		389	1986

Reasons for lack of sulfur dioxide emissions data: Pascagoula, measured, but measurement units not convertible to standardized format; Auburn, Dade County, not measured by plant.

* *Commerce measures all oxides of sulfur, not sulfur dioxide specifically.*

† *90 percent of Lakeland's fuel is high-sulfur coal; only 10 percent is refuse-derived fuel.*

Emission Control Devices		
Electrostatic precipitator (no. of fields)	Fabric filter	Scrubber
	■	■
	■	■
	■	■
	■	■
3		
3		
3		
2		
3		

Emission Control Devices		
Electrostatic precipitator (no. of fields)	Fabric filter	Scrubber
	■	■
	■	■
	■	■
2		
3		
4		
3		
	■	■
3		
5		wet
3		
2		

Table 4-15 Emissions of Oxides of Nitrogen

Plant	Achieve State-of-the-Art Level (100 ppm)	Emissions of Oxides of Nitrogen (ppm)	Year Operations Began	Emission Control Electrostatic precipitators (no. of fields)
Commerce	■	90.2	1987	
Pigeon Point		115	1987	3
Lakeland		184	1983	5
Oswego		197	1986	2
Biddeford/Saco		202	1987	
Baltimore		203	1985	4
Claremont		236	1987	
Westchester		240	1984	3
Marion County		293	1986	
Albany		310	1982	3
Tulsa		367	1986	3

Reasons for lack of oxides of nitrogen emissions data: Pascagoula, Tampa, measured but measurement units not convertible to standardized format; Auburn, Dade County, not measured by plant.
NA, Information not provided by plant.

of variation, and highest levels dramatically exceeded state-of-the-art levels, typical of pollutants that have become a matter for concern relatively recently. However, the newest scrubber technologies regularly attain emissions levels below 30 parts per million for each gas, making the state-of-the-art standard more reliably attainable.

Oxides of Nitrogen

Oxides of nitrogen are produced during combustion when nitrogen-containing garbage, such as yard wastes and food wastes, are burned, and when the nitrogen in air is oxidized during combustion. Of the 11 study plants reporting emissions of oxides of nitrogen (Table 4-15), the only one that achieved the state-of-the-art emissions level

Devices			Other Systems for Reduction of Oxides of Nitrogen Emissions	
Fabric filter	Scrubber	Oxides of nitrogen control devices		Season of Test(s)
■	■	Ammonia injection (Thermal $DeNO_x$)		Spring
			Flue gas recirculation	Winter
	wet			Summer
				Fall
■	■			Spring
				Winter
■	■			Spring
				NA
■	■			Fall
				Winter
				Summer, fall

of 100 parts per million is Commerce. Commerce is also the only plant with an emissions control device specifically designed to chemically reduce oxides of nitrogen. It uses the Thermal De-NO_x selective noncatalytic reduction system that injects ammonia into the furnace to react with and neutralize oxides of nitrogen. Pigeon Point had the next lowest emissions, only slightly above the state-of-the-art level. It has a dual-chambered furnace and a flue gas recirculation system that stabilizes combustion temperatures and slightly lowers the amount of oxygen entering the furnace, thus reducing formation of oxides of nitrogen. The dual-chambered furnace at Oswego probably accounts for that plant's ranking as fourth lowest for emissions of oxides of nitrogen.

Table 4-16 State-of-the-Art Emissions Summary

Plant	Achieved State-of-the-Art Emissions Level for: Particulates	Carbon monoxide	Dioxins / furans	Hydrogen chloride	Sulfur dioxide	Oxides of nitrogen
Commerce	■	■	■	■	■	■
Biddeford/Saco	■		■	■	■	
Marion County	■	■		■		
Claremont	■	■	NC			
Pigeon Point	■	■	NC			
Tulsa	■	NC				
Oswego		■				
Westchester		NA				
Albany		NA				
Baltimore	■	■	NA	NA		
Lakeland		NA	NA	NA		
Tampa		NC	NA	NA		NC
Auburn†		NA	NA	NA	NA	NA
Dade County		NA	NA	NA	NA	NA
Pascagoula		NC	NA	NC	NC	NC

NA, Information not provided by plant.
NC, Measurement not comparable with others.
* *Since 90% of Lakeland's fuel is coal, rather than municipal solid waste or refuse-derived fuels, the state-of-the-art levels identified for garbage-burning plants may not apply.*
† *Closed in 1990.*

The selective noncatalytic reduction system at Commerce is a new system in a new plant; operators were still working to improve its effectiveness at the time of INFORM's visit. The tour guide, Donald Avila, pointed out that the ammonia injection sites had been recently moved in an attempt to find the location in the furnace where the temperature and oxygen conditions are best suited for the conversion of oxides of nitrogen in the combustion gases into nitrogen and water vapor.

Variation in the nitrogen content of the wastes may account for some of the variation in emissions levels of oxides of nitrogen seen in the study plants. Lakeland, for example, may have had comparatively low emissions because only 10 percent of its fuel is solid waste, and much less than half of that is nitrogen-containing yard wastes.

State-of-the-Art for How Many Pollutants? (no./ no. measured)	Emission Control Devices: Electrostatic precipitator (no.of fields)	Fabric filter	Scrubber	Devices for control of oxides of nitrogen	Year Operations Began
6/6		■	■	■	1987
4/6		■	■		1987
3/6		■	■		1986
2/5		■	■		1987
2/5	3				1987
1/5	3				1986
1/5	2				1986
0/5	3				1984
0/5	3				1982
2/4	4				1985
*/3	5		wet		1983
0/2	2				1985
0/1					1981
0/1	3				1982
0/1	2				1985

(Its wet scrubber may also play a role.) Since the amount of yard wastes in the waste stream is greater during and after the growing season, seasonal variations in production of oxides of nitrogen can be expected at individual plants. The data in Table 4-15 do not show any relationship between emissions of oxides of nitrogen and the season of the test when different plants are compared, probably because there are too many other variables affecting the level of the emissions: composition of the garbage in each community, particular composition on the day of the test, furnace characteristics, and test conditions, among them.

Clearly, however, while reduction in the amount of nitrogen-containing wastes entering the furnace may play a role in reducing production of oxides of nitrogen, complete elimination of them is unlikely because oxides of nitrogen are also produced from nitrogen present in the air (approximately 80 percent of air is nitrogen). Thus, both source reduction or composting of food and yard waste and technologies designed specifically for control of oxides of nitrogen are necessary to attain state-of-the-art emissions levels.

Summary of Air Impacts

Table 4-16 summarizes the information about the air impacts of the 15 study plants. It compares which plants measured which pollutants and which achieved state-of-the-art emissions levels with the type of emissions control equipment they have and the year they began operations. It shows:

- Only three of the plants measured all six pollutants for which INFORM established state-of-the-art emissions levels in ways that can be compared (Biddeford/Saco, Commerce, and Marion County); six plants measured five of the six (Albany, Claremont, Oswego, Pigeon Point, Tulsa, and Westchester).
- Only one plant providing comparable data, Commerce, attained state-of-the-art emissions levels for all six of these pollutants.
- One plant providing comparable data (Biddeford/Saco) achieved state-of-the-art levels for four pollutants, and one (Marion County) did so for three pollutants.
- Six plants did not achieve state-of-the-art levels for *any* of the six pollutants: Albany, Auburn, Dade County, Pascagoula, Tampa, and Westchester. (Neither did Lakeland, but state-of-the-art levels identified for garbage-burning plants may not strictly apply to this plant since 90 percent of its fuel is coal.)
- Scrubbers with fabric filters are more effective in reducing air emissions than electrostatic precipitators alone: three of the four plants with this combination of equipment are the plants that achieved state-of-the-art emissions levels for three or more pollutants. While some plants with electrostatic precipitators had state-of-the-art levels of particulate emissions, these devices do not reduce emissions of other pollutants such as acid gases, dioxins/furans, and heavy metals.
- Specific technologies such as selective noncatalytic reduction and flue gas recirculation (along with wet scrubbing and dual-chambered, controlled air furnaces) are essential for reducing emissions of oxides of nitrogen.
- In general, newer plants had lower emissions than plants built only a few years earlier, probably because they tend to have more technologically sophisticated and/or more equipment.

Additionally, a wide variety of plant design and operational features play a role in reducing emissions but, as discussed at the beginning of this "Air Impacts" section, identifying relationships among variables is not easy because of the many variables involved and the diversity of technologies and practices in use.

Ash Impacts

The ash residue left after combustion of municipal solid waste has become the focus of one of the most intense debates over garbage incineration at waste-to-energy plants. Regarding the ash itself, questions include how much ash is acceptable, whether and to what extent this material is toxic, and how it should be tested for the presence and leachability of toxic substances. Regarding how ash should be managed, there are also a variety of issues. State-of-the-art practices involve separating fly and bottom ash; containing ash while it is still in the plant; treating it if possible to minimize toxic impacts (leaching potential, dispersion); transporting it wet in leakproof, covered trucks; and disposing of it in ash-only monofills that have liners to protect groundwater, leachate collection and treatment systems to capture liquid percolating through the ash, groundwater monitoring systems, and daily covering to prevent the ash from blowing around.

INFORM found that none of the 15 plants in this study employ all of these state-of-the-art practices and that ash handling and disposal techniques are not at all standardized.

Scope of the Ash Problem

As the number of garbage-burning plants in the United States increases, not only will the volume of ash requiring disposal increase but also, at the same time, as more efficient air emissions control equipment traps more pollutants, the ash is likely to contain more toxic materials that may potentially be released into the environment. This points to the need to minimize the amounts of pollutant precursors entering the incinerator in the first place, rather than relying on emissions control devices to capture pollutants.

Ash amounts

Although burning garbage does successfully reduce the amount of municipal solid waste, significant quantities of ash still remain. Looking at INFORM's study plants, the 14 plants for which figures were available produced 2684 tons of ash a day, from 10,433 tons of waste burned. The second largest plant in the study, Baltimore, alone generated 639 tons of ash per day, from 2250 tons of waste burned, although this amount is now being reduced by recovering metals and aggregate.

Ash can be measured either by weight or by volume. Weight is most commonly used because this is the standard measurement for municipal solid waste and because trucks weigh in and out of waste-to-energy plants. Volume, however, is the issue for landfills, where capacity is measured in terms of cubic yards. The two types of measurement are not directly comparable because weight and volume of ash residue vary independently depending on the density of the materials in the garbage, the reagents (such as lime) used for air pollution control, the efficiency of combustion, and the amount of water in the ash (water is used to cool the ash when it is removed from the furnace).

Good combustion of garbage leads to large reductions in weight and volume. In Japan, where noncombustible materials (metals, glass) are routinely removed from the

Table 4-17 Ash Amounts

Plant	Weight: Percent of total burned, Reported by plant	Weight: Percent of total burned, Calculated by INFORM*	Weight: Total (tons/day)	Volume (percent of total burned; plant estimate)
Biddeford/Saco	10%	12%	75	NA
Tampa	18-19 (dry)	17	170	5%
Dade County	20	18	308	5
Pigeon Point	20 (dry)	46	180	NA
Marion County	22	24	120-125	5-10
Pascagoula	25	20	25	10
Tulsa	25	25	235	5-7
Westchester	25	26	461	5
Commerce	25	30	100	10
Baltimore	25	28	639	10
Albany	28	28	110	20-25
Claremont	36	36	61	12
Oswego	40	39	75	10
Auburn	50	65 (wet)	120 (wet)	11
Lakeland	NA	NA	NA	NA

NA, Information not provided by plant.

* *By dividing tonnage of ash produced (as reported by plant) by tonnage of solid waste burned (as reported by plant).*

waste stream, the goal is to obtain ash volumes that are 5 percent of the volume burned in incinerators that burn more than 200 tons per day.

The 14 INFORM study plants providing ash weight information reported percentages varying from 10 to 50 percent of the original waste. Recalculating these waste percentages (by dividing the figures provided by the plants for tonnage of ash produced by the figures they provided for tonnage of solid waste burned), INFORM obtained a range of 12 to 65 percent. The plants with the two highest calculated ash weight percentages also had the largest discrepancies between the calculated percentages and the percentage reported by the plants (Auburn: 50 percent reported, 65 percent calculated; Pigeon Point: 20 percent reported, 46 percent calculated). These may relate to differences between wet and dry weights. Leaving these two plants out, calculated ash weight percentages varied from 12 to 39 percent. Both sets of figures are shown in Table 4-17.

Biddeford/Saco reported the lowest ash weight percentage (10 percent), followed by Tampa (19 percent), Dade County (20 percent), and Pigeon Point (20 percent). With the exception of Tampa, all of these are refuse-derived fuel plants. Since some noncombustibles are removed from the fuel for these plants, and since the fuel is homogeneous, combustion efficiency is likely to be high. Nine of the plants reported ash weights falling between 20 and 29 percent of total waste volume, making this the most common range for plants in this study.

Although 12 plants also reported ash volume percentage figures, this information is more impressionistic than the weight data. Since plants generally do not measure either waste or ash volumes, the ash volume percentages reported and shown in Table 4-17 can only be considered as estimates. Five of the 12 plants providing volume percentage estimates reported that information on the total volume of ash produced was unavailable. The volume data provided by the other plants seemed inconsistent. Pascagoula, for example, reported 25 tons per day of ash and 50 cubic yards per day (or 0.5 ton per cubic yard, a low figure), while Baltimore reported 639 tons per day and 675 cubic yards per day (for almost 1 ton per cubic yard, a high figure). (Standard ash weight/volume ratios fall in the range of 1200 to 1500 pounds per cubic yard.) Thus, the reported ash volume percentages, which range from 5 to 25 percent, must be treated with caution.

Making assumptions based on the plants in this study, it is possible to create an order-of-magnitude estimate of the total amount of ash requiring disposal from the United States' 128 operating waste-to-energy plants, although no exact national figure is available. The average weight of ash as a percentage of garbage burned in the study plants is 24 percent (leaving out the two plants with large discrepancies between calculated and reported percentages). Plants in this study operate, on average, at 76 percent of their design capacity. Thus, multiplying the approximate 84,000 tons per day design capacity of the 128 plants in this country by the 76 percent operational level and the 24 percent ash production figure yields a total of more than 15,000 tons per day of ash requiring disposal, or more than 5.5 million tons per year. This is approximately 3 percent of the total municipal solid waste stream. As incineration of municipal solid waste increases, this figure will increase. One study estimates ash amounts of 50,000-55,000 tons per day, or 18 million tons per year, by 1993.[5]

Thus, while it is true that garbage burning leaves only a fraction of the original waste stream, that fraction is a significant amount. It still requires landfill space for disposal and, throughout the country, landfills are being closed, either because they have reached capacity or because toxic substances are leaching from them. It is also becoming increasingly difficult to site new landfills. It is thus essential that the need for landfill space be considered when planning new waste-to-energy plants.

As Table 4-18 shows, several of the study plants are already close to running out of landfill capacity. Specifically, Biddeford/Saco projected reaching its landfill capacity

Table 4-18 Landfill Capacity*

Plant	Year Landfill Will Reach Capacity, Projection	Year Current Ash Disposal Agreement Expires
Albany	1988†	NA
Auburn**	1988	NA
Baltimore	2003	2003
Biddeford/Saco	1991	NA
Claremont	2007	2007
Commerce	1993 (2013 if expanded)	1992
Dade County	2068	On-site landfill; no contract
Lakeland	2013	On-site landfill; no contract
Marion County	1995 (landfill); 1998 (monofill)	NA
Oswego	1988†	County owns landfill; no contract
Pascagoula	2005	2000
Pigeon Point	2006 (expansion planned)	2008
Tampa	2008†	No contract
Tulsa	1990	NA
Westchester	2010	2004

NA, Information not provided by plant.

* *Dates based on plant start-up date and data provided by the plants on years of life of landfills; see plant profiles in Appendix A for additional information.*

† *Expansion planned at time plant reported this information; plant did not respond to request for updated information.*

** *Plant closed in 1990.*

in 1991 and Marion County in 1995, while Albany, Oswego, and Tulsa provided data during INFORM's original research showing they expected to run out of space before 1991; they did not include updated information on this point when responding to INFORM's prepublication follow-up questionnaire. With the exception of Lakeland and Dade County, which have on-site landfills, all will reach capacity by 2010.

Further, as Table 4-18 demonstrates, there are discrepancies for some plants between the year the landfill will reach capacity and the year the current ash disposal agreement expires. Whichever happens first, the plant will have to find a new site to dispose of its ash. With waste-to-energy plants generally operating beyond their 20-

Table 4-19 Ash Testing

Plant	Tests* / Additional Materials Tested For	Testing Frequency
Albany	EP Tox, TCLP / combustibles, metals	June 1987– July 1988
Auburn	NA	NA
Baltimore	EP Tox / chlorinated pesticides	Once since 1986
Biddeford/Saco	EP Tox / all heavy metals, moisture, % solids	Quarterly
Claremont	EP Tox	Monthly
Commerce	EP Tox	Eight times between September 1987 and August 1988
Dade County	EP Tox / organics, moisture	Quarterly
Lakeland	None	None
Marion County	EP Tox	None required beyond initial tests
Oswego	EP Tox, TCLP	"Several" since start-up
Pascagoula	EP Tox	Quarterly
Pigeon Point	Dioxins/furans, metals	Eight times between 1986 and 1988
Tampa	EP Tox	Annual EP Tox tests required; many additional tests in conjunction with ash reuse investigations
Tulsa	EP Tox	None required beyond initial tests
Westchester	EP Tox, TCLP	Twice yearly

NA, Information not provided by plant.

* *EP Tox, the EPA's Extraction Procedure Toxicity Test, tests for arsenic, barium, cadmium, chromium, lead, mercury, selenium, and silver. TCLP, Toxic Characteristics Leaching Procedure.*

year permitted lifespan — for a total of 30 years or so — several of the plants will probably run out of landfill capacity while the incinerators still have years of useful life, assuming no new landfill sites or reuse technologies are used. Finding ways to dispose of ash will become more critical as ash wastes continue to increase and landfill sites continue to fill up.

Table 4-20 Ash Handling, Transportation and Treatment

Plant	Handling: Ash Covered in Plant	Transportation: Containers and/or trucks covered	Containers and/or trucks leakproof	Other
Albany		■		
Auburn		■	■	
Baltimore	■	■		
Biddeford/Saco	■	■	■	
Claremont	■	■	■	
Commerce		■		
Dade County	■			
Lakeland				Conveyors to adjacent landfill
Marion County	■	■	■	
Oswego		■	■	
Pascagoula				
Pigeon Point	■	■		
Tampa		■	■	
Tulsa	■	■	■	
Westchester		■		

NA, Information not provided by plant.

Ash toxicity and testing

Although ash toxicity is a matter of growing concern as both ash amounts and toxic content increase, there is, as discussed in Chapter 3, no agreement about which, if any, of the existing testing procedures — including the EPA's Extraction Procedure Toxicity Test (EP Tox) that assesses levels of eight metals — provide reliable information. The controversy centers on how closely the testing conditions mirror actual conditions. Nevertheless, as Table 4-19 shows, 12 of the plants in this study reported using or having used EP Tox. (Starting in 1991, the EPA is requiring plants to use the Toxic Characteristics Leaching Procedure — TCLP — rather than EP Tox.) Six of the plants provided information on regularly scheduled ash testing (ranging from daily to yearly), with the remainder reporting occasional testing or testing only during the initial start-up period.

Treatment	
Ash treated	Method
■	Mixed with phosphoric acid and lime to immobilize lead
■	Mixed with scrubber sludge and lime to immobilize metals
NA	NA
■	Agglomeration and spray wetting

In theory, ash testing should determine how to dispose of ash. In practice, however, only some states require ash that fails EP Tox tests to be disposed of in a hazardous waste landfill. And, even more noteworthy, some states that require special disposal of ash that fails toxicity tests do not require the testing to make that determination.

Ash Handling, Transportation, and Treatment

The two main concerns of ash management prior to disposal are assuring the safety of workers in the plant and preventing ash from escaping into the environment during removal of the ash from the plant for disposal. Both concerns require that ash be contained at all times both inside and outside the plant, usually by wetting the ash and using enclosed conveyors, and that trucks and containers be leakproof. Additional state-of-the-art management techniques include separating fly and bottom ash so the more toxic fly ash can be treated, and treating ash to immobilize toxic substances. None of the plants in INFORM's study reported separating ash in this way.

Table 4-20 shows that only four plants (Biddeford/Saco, Claremont, Marion County, and Tulsa) reported covering the ash in the plant and transporting it in covered, leakproof containers or trucks. During the plant visits, INFORM observed various incompletely enclosed systems for transporting ash within the facilities: for example, an uncovered, dripping ash truck waiting outside of the Dade County plant to be weighed for its short trip to the on-site ashfill. Only seven plants report covering ash inside the plant, and only seven report using leakproof trucks or containers for transportation, although 12 report covering the trucks or containers. Altogether, only six plants use some combination of containing the ash inside and outside the plant.

Ash Disposal

Minimizing the leaching of toxic metals is one of the keys to safe disposal of ash. State-of-the-art techniques for achieving this, as discussed in Chapter 3, include disposing of the ash in a monofill rather than mixing it with municipal solid waste; using composite liner systems consisting of multiple layers of plastic and clay liners sandwiched between leachate collection and leak detection systems; treating the leachate on-site; and preventing ash dispersion through daily covering and moistening.

At the time the research for this study was carried out, the most stringent ash disposal standards for the states in which INFORM's study plants are located were those mandated by 1988 New York State regulations. They define best practices as a single lining with leachate collection for monofills and a double lining with leachate collection for landfills where ash is disposed of together (codisposed) with municipal solid waste. This standard is required for new facilities. Subsequently, Maine has adopted even more stringent regulations, requiring monofills, double linings, leachate collection, and leak detection for all ash disposal.

Table 4-21 summarizes ash disposal practices in the study plants. Two, Claremont and Marion County, have the most complete containment systems, using all of the state-of-the-art techniques identified by INFORM: monofill, multiple liners, leachate collection, and leachate treatment. (This study did not examine leak detection.) One (Albany) has virtually no protective measures, using neither liners nor leachate collection and treatment, and one (Tulsa) did not provide information on leachate handling. The remaining 11 plants use some combination of containment techniques.

While liners and leachate collection and treatment are generally considered superior ash management techniques, disposal practices are very site-specific and depend on geological conditions and treatment methods. Lakeland, for example, does not have liners for its on-site ashfill, but uses a special treatment that, its managers say, minimizes the leaching by cementing the ash; additionally, 90 percent of its fuel is coal. Thus, the absence of liners in the Lakeland landfill may be less serious than it would be if the ash were untreated or if its fuel contained more municipal solid waste.

Ash treatment is also used at the Claremont plant. Ash is mixed with phosphoric acid and lime to immobilize metals; to date, the ashfill's leachate is mostly salty water.

Table 4-21 Ash Disposal

Plant	Landfill Type: Monofill	Landfill Type: Codisposal with MSW	Number of Liners	Leachate: Collection system	Leachate: Treatment
Albany		■	0		
Auburn	■		1	■	■
Baltimore		■	2	■	NA
Biddeford/Saco	■		1	■	■
Claremont	■		2	■	■
Commerce		■	0	■	■
Dade County	■		1	■	■
Lakeland	■		0*	*	*
Marion County	■		2	■	■†
Oswego		■	2	■	■
Pascagoula	■		1	■	**
Pigeon Point	■		1	■	■
Tampa		■	1	■	■
Tulsa		■	0	NA	NA
Westchester	■		1	■	■

NA, Information not provided by plant.

* *Ash cementation; leachate not formed; runoff collection only.*

† *Leachate diluted with fresh water and then used as a spray irrigant.*

** *Leachate sent to sanitary sewer treatment facility if metals in ash exceed EP Tox levels (tested every 3 months); otherwise, leachate released into surface water.*

However, Claremont's new landfill does, unlike Lakeland's, have two liners and a leachate collection system; these measures may ensure greater long-term protection from contamination of groundwater.

The Commerce plant is another example of site-specific disposal conditions. Ash from this plant is codisposed with municipal solid waste in the Puente Hills canyon, a bedrock landfill without liners. A collection dam and monitoring wells take care of what little leachate may be produced under the dry climate conditions of Los Angeles. This unlined codisposal may be inadequate in another part of the country, but in southern California, where the rainfall is so low and the canyon soil is bedrock, the danger of metals leaching is very low, according to Don Avila of the Los Angeles Sanitation District.[6] However, Richard Denison of the Environmental Defense Fund

Ash Disposal: Profile of a Plant

While Marion County has one of the seemingly most complete containment systems, its ash disposal practices illustrate some of the complexities of ash disposal and the lack of standards. (As INFORM did not examine landfill operating practices for all 15 plants, this closer look at Marion County ash disposal is for the purpose of discussion and not comparison.)

One issue concerns the fact that the ashfill is not covered daily with a layer of soil, suggesting the potential for wind dispersal. While the monofill is fully lined for leachate collection, the ash remains uncovered until the active section of the ashfill is full — maybe up to a few years. The plant's managers claim that the wet ash, containing lime from the plant's scrubber system, solidifies in the landfill, preventing dust from forming and rendering any ash cover unnecessary. INFORM has been unable to learn the effects of a long period of dry weather on the conditions of the Marion County ashfill.

Another issue at Marion County involves leachate treatment. Most of the 15 plants send the leachate to sewage or water treatment plants, but Marion County uses the leachate, diluted with fresh water, for irrigation. This spraying of leachate on adjacent farmland seems to negate the role of the liners in isolating the ash from the environment. Oregon state officials claim that high levels of metals are not a problem, citing abundant rainfall as an explanation for the apparent lack of metal buildup in the soil. Not enough data have been gathered or time elapsed to assess whether any toxic substances present in the leachate will affect the environment adversely over the long term.[9]

challenges the assumption that the bedrock can provide adequate protection from leaching, citing the possibility of liquid traveling through rock fissures.[7]

Commerce's method of disposal has also been questioned because levels of lead and cadmium that are higher than the plant's permit allows have been found in its ash. (These levels are not, however, significantly higher than those at other plants.) According to tabulations using data provided by the Sanitation Districts, the Commerce ash failed the California toxic standards for lead 81 percent of the time, and for cadmium 38 percent of the time.[8]

Other Environmental Impacts

Water use, wastewater handling, and truck traffic to and from waste-to-energy plants can all have environmental impacts on surrounding communities. Water use in the boiler can deplete a community's water supply, lack of appropriate wastewater treatment can leave heavy metals and/or ash in the water, and truck traffic can cause noise, air pollution, and traffic congestion.

Water Use

The plants in this study provided enough information about water use and treatment to obtain a general picture of the types of practices in use, although not enough to compare all the plants to each other. Additionally, INFORM did not ask local authorities in the study plants' communities about any disputes or problems related to the incinerators and their water use.

Table 4-22 summarizes the reported data. Reported amounts of water used range from 1000 gallons per day at the now-closed Auburn plant to 460,000 gallons per day at the Tampa plant; municipal water supplies are generally the source, with additional water being taken from wells, a river, a water treatment plant, and storage ponds.

The destination of wastewater also varies. Six of the plants report treating wastewater or sending it to treatment facilities, three report sending it to customers as steam, and seven release it into the sewer system. Since sewer systems are not designed to handle heavy metals, to the extent that such metals get into wastewater, this does not seem to be an appropriate way to dispose of it. INFORM observed berms around places where rain or washwater could come into contact with ash, thereby directing contaminated water to treatment facilities, at Tulsa and Commerce. Biddeford/Saco and Oswego send heated water, which has been used to cool condensers but has not come into contact with ash, into nearby rivers. It is possible that the change in water temperature could affect aquatic life, especially trout. Only Marion County sends untreated water into a river. Lakeland is unique in that it recirculates treated water for its own use; Westchester reported using untreated wastewater in the plant.

Truck Traffic

The amount of truck traffic to and from a plant basically depends on the size of the plant. Larger plants usually have more trucks delivering municipal solid waste and removing ash. The size of the trucks used also plays a role, but INFORM did not obtain that data. Among the study plants, as seen in Table 4-23, daily truck traffic ranged from a total (waste delivery and ash removal) of 500 trucks at Dade County (the plant with the largest capacity) to 17 delivery and 3 removal trucks at Claremont (the plant with the second smallest capacity). Trucks pass through residential areas approaching four of the plants: Albany, Auburn, Commerce, and Oswego.

Plants that receive hundreds of deliveries of garbage each day need to effectively direct and handle that traffic. The largest plant in this study, Dade County, at the time of INFORM's visit, had an early morning line of trucks waiting, with their motors running, to dispose of their loads. This study did not assess how often this occurs at Dade County or at any other plant, but it is clear that the air impact on a community is increased when trucks are forced to wait to dump their loads. It is thus important to focus attention on waiting times.

Table 4-22 Water Use and Treatment

Plant	Amount of Water Used (gallons/day)	Water Source: Municipal	Water Source: Other
Albany	NA	NA	NA
Auburn	1000	■	
Baltimore	NA	■	
Biddeford/Saco	65,000		River
Claremont	NA	■	
Commerce	300,000	■	
Dade County	NA	■	Wells
Lakeland	NA		Storage ponds
Marion County	NA		Wells
Oswego	18,000		Wells
Pascagoula	NA	■	
Pigeon Point	NA	■	
Tampa	460,000		Water treatment plant
Tulsa	Varies, depending on steam production	■	
Westchester	NA	■	

NA, Information not provided by plant.

Water Destination			
Treatment	Steam to customer	Sewers	Other
			NA
		■	
		■	Harbor
		■	Cooling water to river
			No discharges of water
■			Evaporation of wash-down water
		■	
■			Treated water is recirculated
			River
■			Cooling water to river (untreated)
■	■		
	■	■	
■			
	■	■	
■		■	Reused in plant

Table 4-23 Traffic Impacts

Plant	Number of Trucks/Day: Delivery of waste	Number of Trucks/Day: Removal of ash	Capacity (used) of Plants (tons/day)	Trucks Pass Through Residential Areas
Albany	40	4-6	400	■
Auburn	42-60	7	185	■
Baltimore	400	35	2250	
Biddeford/Saco	85-100	4-5*	607	NA
Claremont	17	3	171	
Commerce	66	8	330	■
Dade County	430†	70†	1700†	
Lakeland	40	0**	390	
Marion County	80-120‡	6-10	510	
Oswego	60	5-7	190	■
Pascagoula	25-30	3	125	
Pigeon Point	250***	2	890***	
Tampa	150	10-15	850	
Tulsa	144-166	17	925	
Westchester	150-250	25	1800	

NA, Information not provided by plant.

* *Seven trucks per day for glass and grit removal.*

† *Figures based on 500 total trucks, with 10-20 delivering waste and 2-3 removing ash on site at one time. Used capacity of 1700 tons per day is during reconstruction process.*

** *Ash disposal on site.*

‡ *Weekdays; 10-30 on weekends.*

*** *Trucks deliver waste to RDF processing plant; used capacity is that of RDF processing facility, not electricity generating facility.*

Workers

Chapter 3 pointed out that, although the Environmental Protection Agency has included the recently promulgated American Society of Mechanical Engineers standards for incinerator worker certification in its New Source Performance Standards, there are no standardized national programs for training incinerator operators and no worker safety standards. Yet, estimating an average of 30 operational employees per plant, there would be some 3800 workers currently operating the United States' existing 128 waste-to-energy plants, and 6000 more will be needed if the 200 new incinerators now being planned go into operation. (The 30 operational workers per plant figure is based on six workers per shift and five shifts for a 7-day week, typical of mid-size incinerators. The plants INFORM studied reported from 12 to 210 workers, but these figures in some cases include administrative and maintenance employees as well as operations staff.)

INFORM found training at the 15 study plants to be mostly on the job, with some additional in-house or vendor training and some use of outside classes. The actual experience of plant employees is quite varied: some have previous boiler experience outside the municipal solid waste field or some engineering degrees or engineering experience, but relatively few, even chief facility operators, have any previous garbage-burning experience. Table 4-24 compares the training and experience information provided by the study plants.

The recent American Society of Mechanical Engineers standards do not move worker training much closer to state of the art. As discussed in Chapter 3, they only provide for certification of senior operators, and the training and education required do not ensure a workforce maximally prepared to operate complex equipment in a changing regulatory environment.

The use of safety equipment is an indication of respect for the environment by workers and plant managers alike. While INFORM's primary purpose in visiting waste-to-energy plants was to examine their environmental performance, practices that preserve the health and safety of workers became a concern when it became clear that such practices not only benefit the workers themselves, but also the general public. If workers are in danger from air pollutants or fugitive ash in their workplace, the general public to a lesser extent may be as well. If workers are injured because of careless plant operation, that same carelessness may lead to environmental impacts on the public. The discussion here of safety practices is based mostly on INFORM's observations during plant visits. We did not obtain sufficiently comprehensive information to compare the plants' performances.

Hardhats are important in areas where there is the possibility of injury from falling objects. Every plant operator reported hardhats to be a requirement inside the plant. In general, workers observed by INFORM were wearing hardhats and INFORM researchers were given hardhats to wear on each plant tour.

Workers do not need eye protection in every work situation. It is appropriate, for

Table 4-24 Worker Training and Experience

Plant	Training: On-the-job	In-house program	Outside classes	Information not provided
Albany				■
Auburn			■	
Baltimore		■		
Biddeford/Saco	■	■	■	
Claremont		■		
Commerce	■		■	
Dade County	■			
Lakeland	■			
Marion County	■		■	
Oswego	■	■		
Pascagoula	■			
Pigeon Point				■
Tampa	■	■		
Tulsa	■		■	
Westchester	■			

NA, Information not provided by plant.

example, inside a refuse-derived fuel processing plant such as Lakeland where there are many small flying particles during waste processing; we were given eye goggles there. At another refuse-derived fuel plant, Biddeford/Saco, no eye protection was noticed, but the refuse-derived fuel processing section was not in operation at the time of INFORM's visit.

Protective devices to avoid inhalation of particulates (respirators), aimed at protecting workers in the ash handling sections of the plants, are not routinely worn.

Actual Experience		
Chief facility operator	Shift supervisor	Control room operator
NA	NA	NA
Boiler manager license	Boiler engineer license	NA
In-plant	In-plant	In-plant
Boiler license, power plant experience	Boiler license, power plant experience	Boiler license, power plant experience
Boiler license	Boiler license	Boiler license
Boiler experience	Boiler experience	Boiler experience
Engineering; in-plant	Engineering; in-plant	Engineering; in-plant
NA	NA	NA
In-plant	In-plant	In-plant
Engineer, ASME course	Steam plant	Steam plant
On-the-job	On-the-job	On-the-job
In-plant	In-plant	In-plant
Manager of coal plant	Boiler and incinerator experience	NA
Waste-to-energy plant experience, power plant training	In-plant	In-plant
Boiler operations	Boiler operations	Boiler operations

At Tampa and Tulsa, for example, we observed workers handling ash without wearing respirators. Nor were respirators in use at the Auburn plant which was filled with combustion fumes at the time of our visit (and none were offered to INFORM).

We noticed no ear protection being used by workers in extremely noisy areas during our plant visits, although plant managers report such protection to be available if wanted.

Notes

1 *Waste Age,* November, 1990. These plants have a design capacity of 84,246 tons per day.

2 Personal communication, Citizens' Clearinghouse on Hazardous Waste, Alexandria, VA, to David Saphire, INFORM, January, 1991.

3 US Environmental Protection Agency, Office of Solid Waste, *Characterization of Municipal Solid Waste in the United States: 1990 Update,* June, 1990.

4 Personal communication, McKay Bay Refuse-to-Energy Facility (Tampa) control room operator to Dr. Maarten de Kadt, INFORM, February, 1988.

5 "Recommendations for Policy and Regulations for Residue from MSW Incineration," Toxic Substance Control Commission, State of Michigan, August, 1988, p. 2.

6 Personal communication, Don Avila, Los Angeles Sanitation District, to Dr. Maarten de Kadt, INFORM, March, 1988.

7 Personal communication, Richard Denison, Environmental Defense Fund, to Dr. Maarten de Kadt, INFORM, August, 1990.

8 *Waste Not,* #40, January 31, 1989.

9 *Waste Not,* #37, January 10, 1989.

Chapter 5: The Economics of Waste-to-Energy Plants

The economics of incineration are complex and variable, with each waste-to-energy plant unique in its individual financial arrangements. Plants can be constructed by private vendors, by states or municipalities, or by some combination; financing has a similar variety of sources. Once the plants are built, they can be owned and operated by governmental authorities or by corporations, or owned publicly and operated under contract by private enterprise. Nationwide, 64 percent of operating waste-to-energy plants are publicly owned and 36 percent privately owned; 40 percent are publicly operated and 60 percent privately operated.[1]

Despite this variation, all waste-to-energy plants operate within the same basic financial framework: they incur costs and obtain revenues. Specifically, costs include construction, financing (interest), operations and maintenance (including labor), and ash disposal. Revenue sources include tip fees for garbage delivered to the plant and sales of energy (electricity or steam) and materials separated from the waste stream. Incinerators may also receive public money directly from taxes or through tax incentives of various sorts.

Municipalities served by garbage-burning plants have agreements with the plant operators regarding such topics as who sets and pays tipping fees and who sets prices for and benefits from the sale of energy and other products. And, as with all forms of waste disposal, the municipal governments establish how citizens pay for incineration: through taxes, separate garbage management fees, or some other method.

While detailed economic analysis is beyond the scope of this book, we can provide a look at some basic economic factors through a compilation of data about construction and operations and maintenance costs, and about garbage tipping fees and other revenues, at INFORM's 15 sample plants. By comparing these figures for those plants that provided them, we obtain a picture of their magnitude and variability.

Costs

The costs of burning garbage in a waste-to-energy plant include both capital and operations and maintenance expenses. Here, to facilitate comparison, we have examined construction expenses not only in absolute amounts but also in terms of cost per ton of design capacity and cost per ton of garbage to be burned over an estimated 30-year useful lifetime. We also evaluated annual operations and maintenance expenses on a per ton basis. While we do not present total expenses, we have a basis for exploring the general costs of constructing and operating garbage-burning plants. Throughout this discussion, the costs associated with more recently built plants provide the best perspective on the likely magnitude of costs of new plants.

True costs are understated for all plants because we have not included the costs of financing or of ash disposal. These costs, while significant expenses, are not presented because INFORM was not able to systematically obtain these figures from the plants. Further, we have not adjusted construction costs for inflation; they are given as of the time of construction. Inflation adjustments are not included both because any adjustments depend on the specific price inflator chosen and because making inflation adjustments did not change the comparative analysis of the plants in this study.

Financing costs depend on the interest rates used at the time a plant was financed. A one percent difference in interest rate has multiple effects on the cost of the project. The impact of the state of the economy, and thus of interest rates, on financing is an important dimension of overall incineration costs, but its analysis is beyond the scope of this book.

Ash management costs also vary dramatically. The Dade County plant, for example, disposes of its ash on site, and therefore has minimal transportation costs and no tipping fee for disposal of the ash. At the Tulsa plant, on the other hand, the company that owns and operates the incinerator does not own the ash disposal site. Its costs therefore include both tipping fees and transportation costs. Furthermore, according to plant officials at the time of INFORM's visit, transportation costs might increase substantially in the future because the nearby ash disposal site was almost exhausted and future sites might be as far as 100 miles away. While the tremendous variability of these ash management costs inhibits quantitative comparison of the plants in this study, these costs are important factors requiring careful analysis when an incinerator project is planned, especially when incineration is compared to other forms of solid waste management.

INFORM did not adjust costs for inflation. There is no specific inflation index for waste-to-energy plants. An examination of indices that might be used indicates that inflation between 1981 (the construction date of the oldest plant) and 1987 (the construction date of the newest plant) was approximately 15 percent. While all of the inflators show similar magnitude and trend from 1981 to 1987, there are differences in these for specific years, and in using them we run the risk of distorting costs for a specific plant. For example, slight differences are obtained when inflating costs using the Producer Price Index for "Finished Goods, Excluding Foods and Energy, Capital Equipment" and the index for "Materials and Components for Construction." Since the cost data presented are intended to give a general picture, we have chosen to present all data in current dollars.

Further, adjusting for inflation would not change the analysis presented. The newest plants have construction costs per ton of design capacity that are more than 300 percent higher than those of the oldest plants, but only a 15 percent increase could be attributed to inflation.

Construction

The cost of constructing the 15 plants in the study ranged from just under $4 million to $239 million. Since the variation in these costs partly reflects variations in plant size,

Table 5-1 Construction Costs and Design Capacity

Plant	Year Operations Began	Construction Costs*	Design Capacity (tons/day)	Cost per Ton of Design Capacity
Biddeford/Saco	1987	$ 67,000,000	607	$ 110,038
Westchester	1984	239,000,000	2250	106,222
Commerce	1987	35,000,000	330	106,060
Claremont	1987	17,900,000	200	89,500
Marion County	1986	47,500,000	550	86,364
Pigeon Point†	1987	50,000,000	600	83,333
Baltimore	1985	170,000,000	2250	75,556
Oswego	1986	14,500,000	200	72,500
Tampa	1985	70,000,000	1000	70,000
Tulsa	1986	76,000,000	1125	67,556
Dade County**	1982	165,000,000	3000	55,000
Pascagoula	1985	6,800,000	150	45,333
Albany	1982	16,000,000	600	26,667
Auburn	1981	3,980,000	200	19,900
Lakeland‡	1983	5,000,000	500	10,000

* *Construction costs at time of construction; not adjusted for inflation.*

† *Costs for Pigeon Point do not include separate resource-derived fuel processing facility ($72,300,000 construction costs for 1000 tons per day capacity, or $72,300 per ton of design capacity).*

** *Costs for Dade County do not include an additional $65,000,000 for reconstruction (additional $21,666 per ton of design capacity).*

‡ *Construction costs for Lakeland include only those associated with the refuse-derived fuel section of the plant, not those for the coal-burning electric power plant.*

comparing the plants requires examining the cost per ton of design capacity to compare the plants. On this basis, as Table 5-1 shows, construction costs ranged from $110,038 per ton of capacity for Biddeford/Saco, one of the two newest plants in this study, to $19,900 per ton for Auburn, the now-closed oldest plant, with no operating emissions control equipment. (The $10,000 per ton cost for Lakeland cannot be directly compared because it applies only to the refuse-derived fuel section of the plant. The plant is primarily a coal-burning electric power facility, and thus much of its total cost is not associated with garbage-burning activities.)

Three plants had a construction cost per ton of design capacity exceeding $100,000: Biddeford/Saco, Commerce, and Westchester. Commerce and Biddeford/Saco are two of the newest plants in the study and Westchester is one of the largest. Geographical differences are evident here, since the Baltimore plant is the same size as the one in Westchester, and newer, but had a construction cost of more than $30,000 a ton less than Westchester. According to Charles Miles of Westchester's Department of Public Works, this difference is entirely a result of differences in local costs of construction,[2] including both land acquisition costs and state construction requirements.

With some exceptions, Table 5-1 shows that construction costs per ton of design capacity have been increasing over time. Some of this is undoubtedly due to more and more sophisticated emissions control equipment, and some is due to inflation.

Given this trend towards increasing construction costs, it is reasonable to assume that construction costs for new incinerators, exclusive of financing costs, will continue to exceed $100,000 per ton of design capacity. A recent detailed analysis of incinerator costs by the Environmental Defense Fund supports this assumption: EDF used a construction cost estimate of $250 million (1990 dollars) for a 2000 ton per day plant.[3] This translates into $125,000 per ton of design capacity. For comparison, if the $110,038 per ton (1987 dollars) construction cost of the Biddeford/Saco plant (the most expensive one in INFORM's study) is inflated by 6 percent per year for three years, the cost in 1990 dollars would be $131,057 per ton.

While the cost per ton of design capacity gives a rough sense of comparative costs, it does not take into account the total amount of garbage each plant will burn over its lifetime. The plants in this study, on average, burn only 76 percent of the garbage stipulated in their design rating, but this percentage varies considerably from plant to plant. Table 5-2 looks at the construction cost per ton of garbage to be burned over a 30-year lifetime, calculated by multiplying the average amount of garbage being processed each day at each plant by 365 days per year and 30 years.

On this basis, the construction cost per ton of garbage to be burned varies from $3.64 for Albany to $12.13 for Westchester, excluding Auburn and Lakeland. Overall, the ranking reflects both the cost per ton of design capacity shown in Table 5-1 and the percentage of design capacity actually being used. The two plants with the highest lifetime costs, Westchester and Pigeon Point, were operating at only 80 and 65 percent of design capacity, respectively. While Westchester had the second highest costs per ton of design capacity in Table 5-1, Pigeon Point ranked sixth there. Similarly, Dade County had been operating at only 57 percent of its design capacity while undergoing reconstruction, accounting for its comparatively higher cost per ton of garbage actually being burned (compared to its cost per ton of design capacity), while Baltimore, operating at 100 percent of its design capacity, had a relatively low cost per ton when the actual amount of garbage to be burned is considered. Auburn, which would have had the lowest cost per ton over 30 years ($1.96), has been excluded because it only operated for 10 years; thus its actual construction costs per ton are $5.89.

For the purpose of evaluating likely construction costs per ton of garbage to be burned for new waste-to-energy plants, it is most useful to look at the costs of recently built facilities. Such plants contain more modern and sophisticated equipment (although none contains all available state-of-the-art equipment) and their costs more closely reflect current dollar values. Biddeford/Saco, Commerce, Claremont, and Pigeon Point, which all began operations in 1987, had costs per ton of design capacity of $110,038, $106,060, $89,500, and $83,333, respectively; and 30-year lifetime costs per ton of garbage actually to be burned of $10.08, $9.68, $9.56, and $11.70, respectively. It bears repeating that these figures are in 1987 dollars, do not include financing costs, ash disposal, or operations and maintenance, and assume that each plant will continue to operate at the same percentage of design capacity throughout a 30-year lifetime.

Operations and Maintenance

Daily operations and maintenance expenses are the other factor involved in calculating the costs of burning garbage in waste-to-energy plants. Table 5-3 ranks the plants that provided operations and maintenance figures according to their cost per ton of garbage actually burned, with Commerce the highest at $41.51 per ton and Tulsa the lowest at $13.33 per ton. The costs for Pigeon Point do not include $44.00 per ton associated with operating and maintaining its separate refuse-derived fuel processing facility; if these costs were added, Pigeon Point would have the highest operations and maintenance costs, $72.09 per ton. The average cost of operating these nine plants (excluding Lakeland) is $24.90 per ton. For comparison, the Environmental Defense Fund's detailed analysis of the economics of a hypothetical 2000 ton per day incinerator operating at 85 percent of design capacity assumed annual operations and maintenance costs of $12 million (in 1990 dollars), or $19.34 per ton of garbage burned.[4]

The high operational costs per ton at Commerce could be due to its extensive emissions control equipment, while the high costs at the two next highest plants, Dade County and Pigeon Point, can be attributed partly to the fact that they were operating at only 57 and 65 percent, respectively, of design capacity. In general, however, operational costs per ton do not exhibit any patterns of relationships with age or size of plant. (Further, different plants may define their operations and maintenance costs somewhat differently.)

Overall Costs

Table 5-4 examines the costs per ton of burning garbage over a plant's lifetime, combining the construction figures from Table 5-2 with the operations and maintenance figures from Table 5-3 (and not including plant financing or ash disposal costs). The ranking reflects that of the other tables. Commerce, with high construction *and* operations and maintenance costs comes in the highest, at $51.19 per ton. Tulsa, on the other hand, with its low operations and maintenance costs, has the lowest overall costs, $20.83 per ton.

Comparing the environmental performance of the study plants with their costs is possible only in a general qualitative way for several reasons. First, as discussed in Chapter 4, it is not possible to directly compare the plants' environmental impacts

Table 5-2 Construction Costs and Lifetime Garbage Burned

Plant	Year Operations Began	Construction Costs*
Westchester	1984	$ 239,000,000
Pigeon Point†	1987	50,000,000
Biddeford\Saco	1987	67,000,000
Commerce	1987	35,000,000
Claremont	1987	17,900,000
Dade County**	1982	165,000,000
Marion County	1986	47,500,000
Tampa	1985	70,000,000
Tulsa	1986	76,000,000
Oswego	1986	14,500,000
Baltimore	1985	170,000,000
Pascagoula	1985	6,800,000
Albany	1982	16,000,000
Lakeland‡	1983	5,000,000
Auburn***	1981	3,980,000

* *Construction costs at time of construction; not adjusted for inflation. These costs do not include financing, ash disposal, or operations and maintenance.*

† *Costs for Pigeon Point do not include separate resource-derived fuel processing facility ($72,300,000 construction costs for 890 tons/day actual garbage handled, or an additional $7.42 per ton over 30 years).*

** *Costs for Dade County do not include additional $65,000,000 for reconstruction, and average garbage burned per day reflects amount plant was burning during reconstruction process. When the reconstruction costs are added in, and the capacity being used in 1990 is used (2600 tons per day, or 87 percent of capacity), the cost per ton of garbage to be burned over 30 years becomes $8.08.*

‡ *Construction costs for Lakeland include only those associated with the refuse-derived fuel section of the plant, not those for the coal-burning electric power plant.*

*** *Auburn closed in 1990, so 30-year cost figures are for comparison only. Actual costs, based on 10 years of operation, are $5.89 per ton of garbage.*

Actual Garbage Burned (average)		Construction Cost per Ton of Garbage to be Burned Over 30 Years
tons/day	% of design capacity	
1800	80	$ 12.13
390	65	11.70
607	100	10.08
330	100	9.68
171	86	9.56
1700	57	8.86
510	93	8.51
850	85	7.52
925	82	7.50
190	95	6.97
2250	100	6.90
125	83	4.96
400	67	3.64
390	78	1.17
185	93	1.96

Table 5-3 Operations and Maintenance Costs

Plant	Year Operations Began	Actual Garbage Burned (average) tons/day	% of design capacity
Commerce	1987	330	100
Dade County*	1982	1700	57
Pigeon Point†	1987	390	65
Oswego	1986	190	95
Auburn	1981	185	93
Pascagoula	1985	125	83
Marion County	1986	510	93
Tampa	1985	850	85
Tulsa	1986	925	82
Lakeland***	1983	390	78
Average (without Lakeland)			

Note: Construction, ash disposal, and financing costs not included.
Reasons for plants not included: Albany, Baltimore, Biddeford/Saco, Claremont, Westchester, operations and maintenance costs not supplied by plant.

* *Costs for Dade County include presorting garbage and reflect operations during reconstruction period.*
† *Costs for Pigeon Point do not include separate resource-derived fuel processing facility ($4-5 million annual operations and maintenance costs, or $31.61 per ton of garbage burned).*
** *Year of costs supplied by Marion County plant was not clear.*
‡ *Costs supplied by Tulsa plant were for 1986.*
*** *Costs for Lakeland include only those associated with the refuse-derived fuel section of the plant, not those for the coal-burning electric power plant.*

because only three plants (Commerce, Biddeford/Saco, and Marion County) report emissions of all six pollutants for which INFORM established state-of-the-art emissions levels in a comparable format. Second, only ten plants provided operations and maintenance figures. Third, for the reasons discussed above, the capital costs provided to INFORM by the plants do not include financing or ash management expenses. Nevertheless, several general observations about environmental performance are possible.

Annual Operations and Maintenance Costs, 1988	Costs per Ton of Garbage Burned
$ 5,000,000	$ 41.51
22,000,000	35.45
4,000,000	28.09
1,600,000	23.07
1,500,000	22.21
960,945	21.06
3,250,000**	17.46
4,500,000	14.50
4,500,000‡	13.33
500,000	3.51
	24.90

Clearly, the costs of equipping a plant with the most up-to-date pollution control devices contribute to the overall construction cost, although INFORM did not obtain a breakdown of these costs. Commerce, the plant with the highest costs per ton of garbage burned (when the costs of processing refuse derived fuel at Pigeon Point are left out), is the only plant to achieve state-of-the-art emissions levels for all six pollutants with established levels. It is also the newest plant and the only plant with the combination of an acid gas scrubber, a fabric filter, and equipment for controlling emissions of oxides of nitrogen.

Table 5-4 Combined Construction and O + M Costs

Plant	Year Operations Began	Costs per Ton of Garbage Burned * Construction (over 30 years)	Operations and maintenance (1988)	Total
Commerce	1987	$ 9.68	$ 41.51	$ 51.19
Dade County†	1982	8.86	35.45	44.31
Pigeon Point**	1987	11.70	28.09	39.79
Oswego	1986	6.97	23.07	30.04
Marion County	1986	8.51	17.46	25.97
Pascagoula	1985	4.96	21.06	26.02
Tampa	1985	7.52	14.50	22.02
Tulsa	1986	7.50	13.33	20.83
Lakeland‡	1983	1.17	3.51	4.68
Auburn***	1981	1.96	22.21	24.17

Reasons for plants not included: Albany, Baltimore, Biddeford/Saco, Claremont, Westchester, operations and maintenance costs not supplied by plant.

* *Construction costs at time of construction; not adusted for inflation. Operations and maintenance costs do not include ash management or financing.*

† *Costs for Dade County reflect operations during reconstruction period and do not include captital costs of reconstruction or increased level of operations after reconstruction.*

** *Costs for Pigeon Point do not include separate resource-derived fuel processing facility (construction, $9.64 per ton; operations and maintenance, $31.61 per ton; total $41.25 per ton).*

‡ *Costs for Lakeland include only those associated with the refuse-derived fuel section, not those for the coal-burning electric power plant.*

*** *Auburn closed in 1990; 30-year and total figures are for comparison only. Actual costs, based on 10 years of operation, would be $28.10.*

A plant with good environmental performance is also likely to have relatively higher operations and maintenance costs because simply having modern equipment is not enough to guarantee low emissions levels; the equipment must also be operated properly. These costs are associated with such factors as having enough well-trained workers and keeping the equipment in optimal working condition. The data INFORM obtained, however, do not permit an analysis of such variables as quality of management and local wage rates. Further, a plant that carefully removes prohibited materials from the waste to be incinerated, while it may operate more efficiently, incurs the costs of this additional step.

Finally, the performance of the Dade County plant after it is fully reconstructed may provide insight into the economic feasibility of retrofitting existing waste-to-energy plants. The high per ton costs shown here are partly a result of the plant being operated at diminished (57 percent) capacity while it underwent rehabilitation (by 1990, operation was up to 87 percent of capacity).

The Citizen's Perspective

From the citizen's point of view, the significant costs of garbage burning are those that he or she must pay, rather than those incurred by the owners and operators of the plant. Citizens pay for waste disposal through taxes and other fees; the exact method of payment varies from community to community. They may also pay for incineration through higher tax rates if tax-exempt financing is used for capital construction of disposal facilities, or if land used for waste management is removed from local tax rolls.

The amount citizens pay varies, depending on the types of waste management in their communities, general levels of costs in their area, and other economic factors. However, a look at one community can provide perspective on overall cost patterns.

In the suburban New York county of Westchester, incineration at the plant profiled in this study is the primary waste management practice (73 percent); up to 22 percent of the municipal solid waste is exported for landfilling, and the remaining 5 percent or more is recycled. An analysis of one Westchester County community, Yorktown, showed that, in 1989, residents paid $92 per ton to the county for its waste management costs, including tipping fees for incineration, and $136 per ton to private carters to collect garbage and deliver it to the nearest county facility. Thus, Yorktown residents paid a total of $228 per ton for garbage management. Since, on average, Westchester residents generate 1 ton of garbage per person per year, this translates into approximately $228 per person per year if we assume Yorktown residents produce garbage at the average countywide rate.[5]

Revenues

The total operating revenues a waste-to-energy plant obtains are a combination of tipping fees charged for bringing garbage to the plant, sales of electricity or steam, and sales of materials (primarily ferrous metals) separated out of the waste stream. In some cases, taxes may also be used to support publicly owned and operated plants.

The plants in this study did not provide INFORM with figures on their total revenues, and many did not supply any data at all on revenues. The waste-to-energy business is a competitive business; facility operators often do not permit public scrutiny of financial information.

In some cases, however, we could obtain information on the tipping fees and on prices for energy and other products. This information is shown in Table 5-5. Because

Table 5-5 Revenues

	Partial Revenues*		
Plant	Tipping fees (per ton)	Electricity (per kw-h)	Steam (per 1000 lb)
Albany	NA	–	NA
Auburn	$47.00 (members) $64.00 (nonmembers) $100.00 (special handling)	–	NA
Baltimore	$33.28 (municipal) $34.68 (commercial)	NA	–
Biddeford/Saco	$4.00 (Biddeford) $8.00–10.00 (others)	NA	–
Claremont	NA	$0.09	–
Commerce	$18.00	$0.08	–
Dade County	$22.00	$0.02	–
Lakeland	$12.00 (city) $16.25 (noncity)	$0.07	–
Marion County	$26.00	$0.06	
Oswego	None	–	$3.20
Pascagoula	$16.83	–	$2.00
Pigeon Point	$37.30	$0.03–0.05	–
Tampa	$18.00 (operator) $58.00 (commercial)	$0.026	–
Tulsa	$21.00	$0.02	$2.75
Westchester	$17.00	NA	–

NA, Information not provided by plant. Albany, Baltimore, Biddeford/Saco, Claremont, and Westchester did not provide O&M figures.

– *Not sold by plant.*

* *Tax revenues not included.*

† *Negative number implies that other revenues are required to operate the plant. Cost figures from Table 5-4 do not include financing or ash management costs.*

** *Auburn closed in 1990. This figure, for comparison purposes only, is based on a hypothetical 30-year life for the plant. The actual difference between the tipping fee and combined construction and operations and maintenance costs, for 10 years of operation, would be $18.90 (members).*

Scrap metal (per ton)	Difference between Tipping Fee and Combined Construction and O&M Costs (Table 5-4) (per ton)†
–	
–	$ 22.83 (members)**
–	NA
–	NA
–	NA
–	– $ 33.19
$40–45	– $ 22.31
–	$ 7.32 (city)
$34	$ 0.03
–	– $ 30.04
–	– $ 9.19
–	– $ 2.49
$0–10	– $ 4.02 (operator)
–	$ 0.17
–	NA

tipping fees are charged on the basis of a ton of garbage, it is possible to compare tipping fee revenues with the combined construction and operations and maintenance costs per ton of garbage calculated in Table 5-4. This comparison, in the last column of Table 5-5, shows that tipping fees alone are generally less than the combined construction and operations and maintenance costs incurred by a plant; thus energy and other sales are necessary if plants are to avoid operating losses. (Further, as pointed out earlier, the costs used here do not include financing or ash management.) While it is known that some waste-to-energy plants not in this study (Warren County, New Jersey; Susanville, California; and Rutland, Vermont, for example) have shown financial deficits, operating losses in general are not likely in view of the guarantees of revenue flows built into the financial designs of garbage-burning plants.

With capital costs in the millions of dollars, owners and operators of waste-to-energy plants, as well as providers of construction financing, use various strategies to assure themselves of adequate flows of revenues. They establish contracts to create long-term markets for the plant's products (steam or electricity and secondary materials) and use both contracts and flow-control ordinances to guarantee the amount of garbage the plant will process.

A **flow-control ordinance** declares garbage the property of the municipality once it is placed out for disposal, enabling the municipality to ensure that the garbage will be used to feed the incinerator. In some cases, contracts called "put or pay" contracts specify absolute levels of tipping fees that municipalities must pay to incinerator operators even if not enough garbage is provided. Such contracts and ordinances provide little incentive for source reduction and may actually divert some of the solid waste stream away from recycling, underlining the importance of planning for source reduction and recycling before designing incinerator capacity. However, particularly as source separation laws become more common, flow control ordinances may also direct recyclables to a materials recovery or other processing facility.

Table 5-6 illustrates the different methods used by the plants in INFORM's study to obtain control over revenues. Of the 14 plants providing information, all but three plants have contracts guaranteeing long-term (15-25 years) delivery of garbage; these three (Commerce, Lakeland, and Oswego) are publicly owned and operated. All but two (Dade County and Lakeland) have contracts for the long-term (15-30 years) sale of either electricity or steam, but the electricity produced at Dade County is purchased by Florida Power and Light, the state's electric utility, and the Lakeland plant is actually a power generating station owned by a utility company. In both cases, electricity sales seem assured. Finally, 10 of the plants report flow-control ordinances.

In all, nine plants have created guaranteed revenue flows at both ends of the waste-to-energy process. They either own the garbage or have long-term garbage delivery contracts, and they have long-term sales contracts for their energy products. The length of the contracts is probably tied to the specific financing agreements for the individual plants, rather than their potential 30-year lifespans.

Table 5-6 Control Over Revenues

Plant	Contracts (time in years) Garbage	Electricity	Steam	Flow Control Ordinance
Albany	NA	—	NA	NA
Auburn*	20	—	20	■
Baltimore	20	25	—	■
Biddeford/Saco	20	20	—	■
Claremont	20	20	—	■
Commerce	None	30	—	■
Dade County	15	None	—	■
Lakeland	None	None	—	
Marion County	20	20	—	■
Oswego	None	—	15	NA
Pascagoula	15	—	30	NA
Pigeon Point	20	NA	—	
Tampa	20	21	—	■
Tulsa	20	20	20	■
Westchester	25	25	—	■

NA, Information not provided by plant.
■ *Plant has feature.*
— *Not sold by plant.*
* *Closed in 1990.*

Use of Public Funds

The financial arrangements developed at the time of planning for the construction of waste-to-energy plants can also include a variety of tax and other government incentives. Governments can favor construction of garbage-burning plants through bond guarantees, grants, tax exemptions to bond holders, and removal of real estate from tax rolls. These uses of public money affect the overall cost of the project and can reduce the amount of debt service and taxes that the company operating the facility must pay.

Governments do not generally make comparable financial advantages available to either source reduction or recycling programs, despite the fact that, as a disposal option, incineration ranks third in the widely accepted waste management hierarchy. The Environmental Defense Fund, in its comparison of the economics of incineration and recycling, has called for "leveling the playing field" so that recycling programs have "equitable access to public funds."[6]

Notes

1 *Waste Age*, November, 1990.

2 Personal communication, Charles Miles, Westchester Department of Public Works, to Dr. Maarten de Kadt, INFORM.

3 Environmental Defense Fund (Richard A. Denison and John Ruston, eds.), *Recycling & Incineration: Evaluating the Choices*, Island Press, Washington DC, 1990, pp. 117-118.

4 *Ibid.*, p. 119

5 Maarten de Kadt, "Managing Westchester's Garbage: Building on Experience," *Westchester Environment*, Summer 1990, Federated Conservationists of Westchester County.

6 Environmental Defense Fund, *op. cit.*, pp.163-167.

Chapter 6: The Regulatory Environment

Until recently, regulation of the operation of waste-to-energy plants involved a complex patchwork of federal and state standards and individual permit conditions that were occasionally at odds, frequently confusing, and constantly changing. Investigating the regulations applying to the 15 plants in this study, INFORM found no nationwide consistency. Among the key regulatory questions are: What needs to be regulated? How should pollutants be regulated? How should regulations be enforced? Further, INFORM found that the lack of standardization and the inaccessibility of information (particularly about air regulations) that may be located in a multiplicity of laws and agencies inhibit citizen understanding and comparison.

This picture is now beginning to change, with the new regulations for new and existing incinerators promulgated by the United States Environmental Protection Agency (EPA) in February 1991. The Standards of Performance for New Stationary Sources (Municipal Combustors) (often called the New Source Performance Standards, or NSPS), for the first time on a national basis, establish overall emissions standards for incinerators (standards previously existed only for particulates), and define guidelines for good combustion practices. However, by late 1991, the EPA is required to revise these new regulations to meet more stringent criteria mandated by the 1990 Amendments to the Clean Air Act.

While these new federal regulations introduce some standardization into the existing regulatory patchwork, differences can still exist since federal requirements can be superseded by more stringent state regulations. The research for this study was conducted before such federal standards existed.

The information gathered here on regulations applying to the plants in this study partially explains many of the wide differences observed in incinerator practices and emissions. Until recently, almost nothing had been required of such plants on a national basis; on the state level, permit conditions are generally determined on a case-by-case basis. The fact that none of the 15 plants met INFORM's state-of-the-art standards reflects, in large measure, the absence of any requirement that they do so.

The comparison of federal and state regulations to state-of-the-art standards also sets a context for understanding practices at plants not included in this study. Further, it gives citizens, communities, and government officials a basis for assessing requirements for proposed incinerators and for considering future regulatory needs.

Air Regulations

Current standards for air emissions from garbage-burning plants exist at three levels: federal, state, and plant permit. Before the new EPA regulations, the main federal standards applicable to solid waste incinerators were general ambient air concentration limits for pollutants (discussed below). The only specific regulations for emissions from solid waste incinerators applied to their particulate emissions. In the absence of detailed national incinerator emission standards, some states began to require individual plants to meet specific, but constantly evolving, emissions limits. These state regulations, along with applicable federal standards, were then used, case-by-case, based on local conditions such as terrain and population, to set the operating conditions for a new plant in its permit.

Federal Ambient Air Regulations

The EPA standards for ambient air concentrations of pollutants (the only federal legislation affecting air emissions from waste-to-energy plants before the 1991 New Source Performance Standards) were set under provisions of the Clean Air Act that were aimed at protecting public health. The country is divided into air quality control regions so that local air quality problems can be addressed. The act then requires states to develop programs so that each air quality region meets these standards. In this way, this federal legislation regulates the overall ambient impact of all pollution sources on air quality, not specific emissions from specific sources.

To date, the EPA has set maximum allowable ambient air concentration limits for six so-called **criteria pollutants:** ozone, sulfur dioxide, oxides of nitrogen, carbon monoxide, lead, and particulates. It has also established procedures that proposed new and modified projects must follow to obtain construction permits if they will emit these or any other regulated pollutants in specified quantities. These procedures lead to different emissions requirements depending on whether the air quality control region in which the project is located meets, or fails to meet, the standards for one or more of the criteria pollutants. (Regions that meet a standard are in **attainment** for that pollutant. Ones that do not are in **nonattainment** for that pollutant.) Proposed solid waste incinerators must comply with EPA requirements if they meet a variety of capacity and emissions criteria.

Regions that are in attainment for specific criteria pollutants are allowed to increase pollutant levels by specified increments under the Prevention of Significant Deterioration program. A proposed new source in such a region must conduct an air quality impact review (including modeling and on-site monitoring) to demonstrate that its predicted air emissions levels will not put the region in violation of the attainment standards or exceed the permissible increments. If its proposed emissions levels are deemed acceptable, they usually become the limits set forth in the facility's operating permit.

A proposed plant applying for a Prevention of Significant Deterioration permit must conduct a **best available control technology (BACT)** analysis to propose a control technology for each regulated pollutant. The BACT analysis is done on a case-by-case basis and includes an evaluation of the energy, environmental, and economic costs of alternative control technologies, as well as an assessment of the benefits of reduced emissions. The BACT policy, which includes cost considerations, is flexible, allowing regulators to continually redefine what is an attainable and enforceable emissions standard. In contrast to regulations that set absolute standards, the BACT policy enables states to establish more stringent requirements in individual plant permit proceedings. Finally, the policy also gives manufacturers of pollution control equipment an incentive to invent better control technologies because they can anticipate their new products becoming BACT.

Individual regions may define BACT differently depending on the air quality problems specific to the region. However, to promote some consistency in determining what is BACT, in 1987 the EPA issued guidelines requiring reviewing authorities to consider a dry scrubber and a fabric filter or electrostatic precipitator as BACT for sulfur dioxide and particulate emissions, and combustion controls as BACT for carbon monoxide. Except for a 50 parts per million standard for carbon monoxide, the EPA did not set specific emissions limitations in these guidelines.

If a proposed plant is in a nonattainment region, it must install emissions control technology that allows it to meet the **lowest achievable emissions rate (LAER)** for the nonattainment pollutants. Unlike BACT, LAER requirements do not consider economic, energy, or other environmental factors, just emissions levels. In practice, if a LAER technology is selected as BACT, they are treated as essentially the same, even though in theory they are quite different.

In addition to observing LAER standards, the planners of new emission sources in nonattainment regions must obtain offsets of the nonattainment pollutants through corresponding emissions reductions from other sources in the region. The amount of offsets required varies depending on the severity of nonattainment in the area. The offset system is designed to allow new sources to come on line in nonattainment regions without increasing the region's overall pollution levels.

Finally, while the Environmental Protection Agency did not generally set specific emissions levels for specific types of facilities, it did so for particulate emissions from solid waste incinerators: 0.08 grains per dry standard cubic foot for plants burning less than 250 tons of garbage per day, 0.1 pounds per British Thermal Unit (roughly equivalent to 0.04-0.05 grains per dry standard cubic foot) for plants burning more. These limits are five to eight times higher than INFORM's state-of-the-art level of 0.010 grains per dry standard cubic foot. (INFORM's criteria for identifying state-of-the-art emissions levels, all of which have been achieved in actual operating plants, were discussed in Chapter 3.)

New Federal Incineration Regulations

The Environmental Protection Agency's New Source Performance Standards, proposed in late 1989 and promulgated in February 1991, were the first (beyond the earlier attention to particulate emissions) to broadly and specifically address the performance of municipal solid waste incinerators. The standards apply to incinerators with individual combustion units that have the capacity to burn more than 250 tons per day of municipal solid waste and for which construction, modification, or reconstruction began after December 20, 1989. Existing incinerators of this capacity are covered by different, less stringent guidelines, and smaller incinerators will be the subject of future regulations. As a result of the 1990 Amendments to the Clean Air Act, the EPA is required to revise many of the specific provisions of these new regulations by November 1991.

The new regulations set standards in four basic areas: good combustion practices, emissions levels for six pollutants, monitoring requirements, and operator training and certification. (The original proposal also contained a materials separation requirement recommended by the EPA, but this was deleted in the final version.) They apply, with some variations, both to new and existing garbage-burning plants. As with the current ambient air standards, states must conform at least to the federal standards but are free to adopt more stringent ones. Table 6-1 summarizes the new EPA regulations and contrasts the emissions standards to the state-of-the-art standards identified by INFORM.

For good combustion practices, the new standards set carbon monoxide emissions levels of 50 to 150 parts per million (depending on the specific furnace technology) as evidence of efficient combustion. The upper limit of this range is three times INFORM's 50 parts per million state-of-the-art level — a level regularly attained or even exceeded by new state-of-the-art facilities. The combustion practices section also specifies that incinerators must establish maximum load levels and maximum flue gas temperatures at the inlet for the final particulate control device. These levels then become the plant-specific operating requirements for the incinerator.

The regulations also establish maximum emissions levels or removal efficiencies for particulates, dioxins/furans, hydrogen chloride, sulfur dioxide, and oxides of nitrogen. In some cases, in an attempt to minimize the economic impact on smaller and existing plants, the emissions requirements are less stringent than those for new and larger ones. Additionally, the requirements for some pollutants allow plants to achieve a specified percentage reduction in emissions, rather than an absolute emissions level, whichever is less stringent. While the regulators do not mandate the use of specific emissions control equipment, they do indicate a technological basis for the emissions levels. For example, they indicate that the particulate and acid gas emissions levels for new, large plants can be achieved using a combination of a fabric filter and a spray dry scrubber.

Most of the emissions limits — even the most stringent ones — are less stringent than the state-of-the-art levels used by INFORM. As discussed above, the maximum carbon monoxide level is three times INFORM's state-of-the-art standard. The particulate level, 0.015 grains per dry standard cubic foot, is 50 percent higher than INFORM's 0.010 grains per dry standard cubic foot state-of-the-art level and the lower levels routinely achieved by new state-of-the-art plants. The 180 parts per million oxides of nitrogen limit is greater than INFORM's 100 parts per million state-of-the-art level. For dioxins and furans, the regulations set limits for total dioxin/furan emissions, rather than Eadon toxic equivalents which give a comparative indication of dioxin and furan toxicity levels. It is thus not possible to directly compare them to INFORM's state-of-the-art standards. Further, the use of total dioxin/furan emissions ignores the fact that the toxicity of specific dioxin and furan compounds varies.

Table 6-1 Key Features of New Federal Municipal Solid Waste Incinerator Regulations (NSPS), Compared to INFORM State-of-the-Art Standards

Incinerators Subject to New Regulations

- Burn municipal solid waste or refuse-derived fuel
- Contain individual combustion units with the capacity to burn 250 tons per day or more
- Construction, modification, or reconstruction began after December 20, 1989

New Source Performance Standards	INFORM State-of-the-Art Standard
Materials Separation	
None	Recyclables, noncombustibles, and wastes containing toxic materials or pollutant precursors removed
Good Combustion Practices	
Carbon monoxide emissions: 50-150 parts per million (depending on furnace type)	50 parts per million
Plant-specific maximum load level	
Plant-specific maximum flue gas temperature at inlet to final particulate control device	
Pollutant Emissions Levels (7% O_2, dry basis)	
Particulates	
0.015 grains per dry standard cubic foot	0.010 grains per dry standard cubic foot
Dioxins/furans	
30 nanograms per dry standard cubic meter — total dioxins and furans	0.10 nanograms per dry normal cubic meter — Eadon toxic equivalents
Sulfur dioxide	
80% reduction, or 30 parts per million (whichever is less stringent)	30 parts per million
Hydrogen chloride	
95% reduction, or 25 parts per million (whichever is less stringent)	25 parts per million
Oxides of nitrogen	
180 parts per million	100 parts per million

(continued on next page)

New Source Performance Standards	INFORM State-of-the-Art Standard
Pollutant Emissions Levels (cont'd.)	
Heavy metals	
No individual standards; particulate emissions as surrogate	Not defined; further research needed to identify lowest regularly attainable emissions levels
Monitoring Requirements	
Continuous monitoring	
Carbon monoxide, opacity, sulfur dioxide, oxides of nitrogen	Furnace and flue gas temperature, steam pressure and flow, oxygen, carbon monoxide, opacity, sulfur dioxide, oxides of nitrogen
Annual stack tests	
Particulates, dioxins/furans, hydrogen chloride	Particulates, dioxins/furans, hydrogen chloride, metals
Operator Training and Certification	
American Society of Mechanical Engineers certification standards for chief facility operators and shift supervisors	Formal academic and practical education; supervised on-the-job training; formal testing; periodic reevaluation

Only the hydrogen chloride and sulfur dioxide emissions limits appear to match INFORM's state-of-the-art levels: 25 parts per million and 30 parts per million, respectively. However, the new regulations allow these absolute levels to be ignored if hydrogen chloride emissions are reduced by 95 percent and sulfur dioxide levels by 80 percent (comparing the concentration in the flue gas before it passes through emissions control devices with the emitted concentration). For moderately sized plants which could have flue gas concentrations of 1000 parts per million for hydrogen chloride and 300 parts per million for sulfur dioxide, these percentage reduction requirements would permit emissions of 50 parts per million and 60 parts per million, respectively.

In the new standards, the Environmental Protection Agency has used the particulate emissions standard as a surrogate for emissions of heavy metals (such as lead, cadmium, and mercury). That is, plants are not required to measure heavy metal emissions directly and will be considered to have acceptable levels of heavy metal emissions if their particulate levels fall within the established standards (heavy metals in flue gas can condense onto particulates).

However, tests have shown that there can be mercury emissions even with good particulate collection, since mercury volatilizes at relatively low temperatures and may exist mainly in the gas phase. Therefore, control of mercury emissions is not achieved solely by particulate control devices or by low temperatures of the flue gas entering the particulate collection device. It also depends in part on removal of mercury from the waste stream and on new control devices such as sodium sulfide and activated carbon injection and activated carbon beds that are being tested in Europe.

The 1990 Amendments to the Clean Air Act called for numerical limits for mercury, cadmium, and lead emissions from municipal solid waste incinerators. Such standards are scheduled to be developed and added to the new federal regulations by late 1991.

The regulations also establish monitoring requirements. For new plants with individual combustion unit capacities of 250 tons per day or more, these include continuous monitoring for oxides of nitrogen, sulfur dioxide, opacity, and carbon monoxide, and annual stack tests for particulate matter, dioxins/furans, and hydrogen chloride. Requirements for existing plants vary.

Finally, the EPA regulations specify that chief facility operators and shift supervisors be certified according to standards promulgated by the American Society of Mechanical Engineers. As discussed at greater length in Chapter 3, these certification standards, which include no formal training component, fall far short of state-of-the-art training and certification programs identified by INFORM.

The EPA's original proposal included materials separation standards requiring a 25 percent reduction in solid waste entering the incinerator. As with recycling regulations more generally, the base from which the 25 percent reduction was to be calculated was ill-defined. Nevertheless, the reduction could have come from some combination of paper, metal, glass, and plastic (which could have been recycled through incorporation into marketed products), or yard waste (which could have been composted). In any case, the removal of noncombustible and recyclable material, leaving only soiled combustibles, would improve plant performance by improving the homogeneity and combustibility of the waste stream, thereby reducing the air emissions and the quantity and toxic content of the ash.

This proposal represented a new direction for the EPA, an indication that it intended to actively support its solid waste management hierarchy that places waste reduction and recycling ahead of incineration and landfilling. However, in response to administration and incinerator industry pressure, the EPA removed the materials separation requirement from the final version of the regulations.

State Air Regulations and Permit Conditions

In the absence of specific federal regulations, with increasing numbers of waste-to-energy plants coming on line, with growing public concern about the health impact of these incinerators, and with the development of ever more sophisticated control equipment and operating practices, many states have moved beyond the federal ambient air regulations to develop more stringent, specific, or extensive standards. These standards included requiring BACT or LAER for all new sources, regardless of size; setting performance standards such as stack emissions levels for some criteria pollutants, in addition to federal technological or ambient air standards; considering a variety of pollutants other than the six criteria pollutants; and mandating monitoring techniques, testing and reporting schedules, and operator certification. While the

EPA's New Source Performance Standards establish federal regulations in a variety of areas where they did not exist before, state regulations will continue to differ from federal standards since states still have the right to institute more stringent requirements than those nationally mandated. Some, like those in New York State, are already more stringent than NSPS.

INFORM examined state regulations and local permit conditions in the 10 states in which the study plants were located to discover any general regulatory patterns. The review revealed widespread variation in the use of regulatory strategies. These basic strategies and the variability of their use (rather than specific details of the regulations) are highlighted here. The regulatory limits included here were in effect when research for this study was carried out in 1988.

Table 6-2 shows that three states established numerical stack emissions limits for the criteria pollutants for waste-to-energy plants: California, for particulates, sulfur dioxide, oxides of nitrogen, and carbon monoxide; Maryland, for particulates, sulfur dioxide, and carbon monoxide; and New York, for particulates. Some of these limits are less stringent than INFORM's state-of-the-art levels and the new EPA standards. However, the particulate limits in New York and California are more stringent than the new federal ones.

The seven states that did not establish regulatory limits use case-by-case BACT reviews to set emissions levels in individual plant permits and thus can make permit levels more stringent as better technology develops. Plant permits in Maine and Florida, for example, have set carbon monoxide emissions limits of 100 parts per million, considerably below California's 400 parts per million limit, set in 1984, but still well above INFORM's state-of-the-art level of 50 parts per million.

Several of the states adopted more stringent pollutant permitting programs than those established by federal criteria. Maine and Oregon require BACT review of all sources, regardless of size. California, New York, Maryland, and Florida have stricter ambient air concentration levels that trigger use of LAER than federal standards.

The relationship between a state's air regulations and its specific needs partially explains the variation in regulations across the country. Densely populated states with higher solid waste generation rates and/or poor air quality, such as New York, California, and Maryland, are more likely to standardize permit conditions with state regulations. States like Mississippi and Oregon, on the other hand, which have only one large garbage-burning facility and no current plans for others, may see little need to look beyond the federal standards in their case-by-case permitting.

Many states also examine emissions of pollutants other than the six criteria pollutants during the permitting process, comparing emissions levels and ambient air quality predicted by the permit applicant with levels established by a health risk analysis. These pollutants include a diverse group of metals, acid gases, and products of incomplete combustion, some of which are included in the new EPA incinerator standards. The pollutants of concern vary considerably from state to state and are expected to change over the years as some pollutants, probably metals, are added. In

Table 6-2 State Regulation of Criteria Pollutants (1988)*

Pollutant	INFORM'S State-of-the-Art Emissions Levels	California	Maryland	New York	New EPA Incinerator Standards
Particulates (grains per dry standard cubic foot)	0.010	0.010	0.015	0.010	0.015
Sulfur dioxide (parts per million)	30	30	40	BACT	30 (or 80% reduction)
Oxides of nitrogen (parts per million)	100	140-200	BACT	LAER/BACT	180
Carbon monoxide (parts per million)	50	400	100	BACT	50-150

* *States listed are those in INFORM's study that had set state regulations. BACT review was used in other states and for other criteria pollutants.*

most states, regulators use lists of these pollutants as guidelines for what to consider during permitting: however, only two of the states examined here (New York and Oklahoma) explicitly require testing for the pollutants on these lists. And, once the tests involved in the permitting process are completed, only four states in this study (California, Delaware, Maine, and New York) require periodic compliance testing for any of the pollutants. (The new EPA regulations will alter this since they require ongoing testing of emissions levels.)

At the time of this study, two noncriteria pollutants were of the most concern at the state level: hydrogen chloride and dioxins/furans, both of which are included in the new EPA regulations. Eight of the ten states (all except Delaware and Mississippi) set state or permit levels for hydrogen chloride, and five set limits for dioxins and furans (New Hampshire, New York, Oklahoma, Maine, and Oregon — dioxins only).

However, the variety of units and measurement conditions used in stating these levels makes comparison difficult and in some cases impossible. Many dioxin and furan limits, for example, use units of pounds per hour, which are dependent on the size of the plant, making interplant comparison impossible without gas flow rate data. Only New York expresses these limits in a size-independent way, as a concentration of toxic equivalents; thus, only its standard (0.2 nanograms per normal cubic meter) can be compared to INFORM's state-of-the-art standard (0.10 nanograms per normal cubic meter). (The federal regulations are expressed as concentration of total dioxins and furans and cannot be compared to these standards.) Similarly, hydrogen chloride limits, while mostly but not always expressed in parts per million, are measured over a variety of **averaging times** (the amount of time over which emissions are measured and then averaged to obtain a per-hour figure). Those limits that are expressed in parts per million are less stringent than INFORM's state-of-the-art level and the new EPA standards. This difficulty of comparing information from different states is one of the findings of this study. Table 6-3 summarizes the noncriteria pollutant levels.

States also mandate monitoring techniques. With the exception of Mississippi and Oregon, all the states in this study specify continuous emissions and process monitors that plant design must include, some through existing or proposed state regulations and some through individual permits. The monitoring required varies from eight factors for New York (opacity, furnace temperature, carbon monoxide, sulfur dioxide, oxides of nitrogen, oxygen, carbon dioxide, and combustion efficiency) to two for Oklahoma (carbon monoxide and oxygen). Since some of the technologies for continuously monitoring air emissions are still undergoing change (that for hydrogen chloride, for example), these state requirements are likely to evolve further.

Finally, enforcement methods also vary from state to state. Five states (California, New York, Delaware, Florida, and Oregon) require periodic stack emissions testing although others may require testing only if the regulatory agency suspects permits may be exceeded. All except Mississippi mandate reporting of data from the continuous emissions or process monitors at specified intervals, and Maine requires continuous automatic transmission of data to the State Department of Environmental Protection (telemetering). This process gives the state real data on a timely basis when limits are exceeded instead of forcing regulators to simply suspect noncompliance.

Issues of Air Regulation

This brief survey of air regulations highlights the key issues that faced regulators trying to set permit conditions for waste-to-energy plants and citizens trying to understand these conditions before the EPA's national municipal waste combustion regulations took effect: complexity, lack of standardization, and inaccessibility of information. INFORM carried out this 15-plant study against the backdrop of this regulatory disarray, and this confusion will continue to exist until national standards are completely in place, a process that will take place over a period of several years. (Some regulatory differences will continue to be inevitable since states still have the authority to impose regulations stricter than federal standards.) Further, while the new regulations introduce some much needed standardization, other issues remain, such as how emissions limits should be determined and expressed, how to update regulations as new technologies are demonstrated, and what are appropriate averaging times.

Table 6-3 State or Permit Limits for Selected Noncriteria Pollutants

State	Hydrogen Chloride	Dioxins (polychlorinated dibenzodioxins)	Furans (polychlorinated dibenzofurans)
INFORM's State-of-the Art Standards	25 ppm	0.10 ng/Nm^3 (Eadon toxic equivalents)	
California	30 ppm	No limit	No limit
Florida	0.127 lb/MBTU	No limit	No limit
Maine	30 ppm	3.3 x 10^{-6} lb/h	1.0 x 10^{-6} lb/h
Maryland	50 ppm	No limit	No limit
New Hampshire	50 ppm	3.4 x 10^{-7} lb/h	4.75 x 10^{-6} lb/h
New York	50 ppm	0.2 ng/dsm^3 (Eadon toxic equivalents)	
Oklahoma	0.5 ppm, ambient	1.0 x 10^{-6} lb/h	1.0 x 10^{-6} lb/h
Oregon	10.5 lb/h	1.70 x 10^{-6} lb/h	No limit
EPA's New Incinerator Standards	25 ppm*	30 ng/Nm^3 (total, not toxic equivalents)	

* *Or 95% reduction, whichever is less stringent.*

Inaccessibility of Information

INFORM found that the inaccessibility of information about air regulations is one of the most significant factors limiting serious discussion of air pollution issues and resource recovery standards. Without clear, easily available information, presented in a standard format, it is difficult, and in some cases impossible, to compare requirements and environmental performance at individual incinerators to state-of-the-art standards.

Information is inaccessible for several reasons.

- Information about regulations is scattered among individual plants, local regulatory agencies, state agencies, and regional EPA offices.
- Specific types of information (about air emissions, for example) are located in different agencies in different states.
- Current regulations are sometimes hard to locate because they are still in draft form, or are undergoing revision and review.

- Air regulations are typically expressed in specialized legal and scientific terminology that can differ from state to state.
- Emissions levels, as discussed below, are determined using a variety of measurement standards and methods and expressed in a variety of units with differing correction factors; converting them into a uniform format for comparison is often impossible since other plant data needed to make the conversion may be unavailable. (Appendix B explains the conversions used in this study.)

Lack of Standardization

The overlapping and inconsistent nature of federal ambient air regulations, state regulations, and permit-setting has clearly been one of the main contributors to this informational confusion, causing permit conditions to vary from state to state and from plant to plant. Different pollutants have been regulated, different techniques for managing pollution have been used, and different acceptable emissions levels have been set.

But the lack of standardization goes beyond this: there has also been no agreement about the way emissions levels should be expressed.

- Emissions limits can be expressed in units representing air concentrations (parts per million or grains per dry standard cubic foot); weight per unit of time (pounds per hour); or weight per amount of fuel burned (pounds per ton of fuel or pounds per million BTU), thus inhibiting comparison among plants. Conversion to common denominators is not always possible, given the available data.
- Plants are sampled under different conditions that affect emissions measurements, such as temperature, moisture, and percent of carbon dioxide and oxygen. Thus, correction factors must be applied so measurements are expressed for comparable conditions. Without a defined set of "standard" conditions to which actual measurements could be corrected, accurate comparison among plants will remain difficult.
- The averaging time — the standard amount of time over which emissions are measured, again for purposes of fair comparison — is critical. Actual emissions levels are constantly varying due to variation in waste composition and combustion conditions. Established averaging periods can be shorter (1 to 6 hours) or longer (up to 30 days). Shorter averaging times are better suited to controlling the number and severity of exceedences (emissions greater than the permitted level) since a given exceedence represents a greater percentage of the averaging time when the time is shorter.

  For example, consider a plant that emits 500 parts per million of some pollutant one hour, but only 50 parts per million for each of the next nine hours. If the emissions limit is 100 parts per million, the average emissions level is five times its permitted level for the first hour, and 2.75 times the permit level averaged over the first two hours. If a short-term, 2-hour averaging time were applicable, the plant would exceed its limit for the first two hours. But if averaged over a period of 10 hours, the hourly level falls to 95 parts per million, within the permit conditions even though the plant emitted five times the absolute permit level over 1 hour. The averaging periods for emissions of oxides of nitrogen and sulfur dioxide at the Commerce plant are currently being disputed by local community groups and the California Air Resources Board.

Remaining Issues

The new national standards establish stack emissions limits as the method for regulating six air pollutants of concern. In doing so, they define the measurement criteria to be used: units, conditions, and averaging times. As adopted, they will make it easier to compare the performance of individual waste-to-energy plants to national standards and to the performance of other facilities, assuming the data are publicly available.

While the benefits of having national incineration standards are clear, however, the content of those regulations is critical too. As mentioned above, some of the specific components of the EPA proposal fall short of INFORM's state-of-the-art standards. Remaining issues and topics of debate depend on the extent to which the public, and state governments, want the maximum protection that state-of-the-art plant operations could provide. The issues involve planning, air and ash standards, and worker training as well.

Most dramatically and unfortunately, the removal of the proposed materials separation requirement from the final regulations took away the one significant requirement that would have put incineration in the context of the solid waste management hierarchy that identifies source reduction and recycling as measures to be used before disposal options. Requiring 25 percent separation in communities planning waste-to-energy plants would have increased the likelihood of attaining EPA's stated national goal of 25 percent source reduction/recycling/composting by 1992. Further, depending on the specific materials removed, separation of recyclable components from the waste stream would have had great potential for reducing both air emissions and the quantity and toxicity of ash.

The EPA's goal was, however, imprecise, in that it neither distinguished specific goals for each of these three techniques nor identified any baseline from which to measure source reduction or recycling. As discussed in Chapter 3, a state-of-the-art garbage management strategy would establish individual source reduction and recycling goals for different materials before designing an incinerator.

Another remaining issue involves emissions limits set. Most of the proposed emissions limits are less stringent than those regularly attainable with state-of-the-art equipment, which will spur further debate. Questions also remain about how emissions limits should be established, and whether regulating emissions by setting stack limits is adequate.

- Should regulators consider what is coming out of the stack rather than the impact of these emissions on the environment or on human health? That is, since impacts are so difficult to assess, should emissions be reduced regardless of their impact? How can the costs of these reduction be considered?

- Should health risk assessments be used to set limits? Health risk assessments for air pollution are riddled with uncertainty and complications: translating animal tests to human health predictions is problematic; the effect of small dosages over long periods of time, as opposed to the large dosages over short periods on which animal studies are based, is unknown; pollutants can enter the body through ingestion of food and water from various sources and by skin contact as well as through inhalation; and combinations of pollutants can have complex and possibly amplified effects, rather than simple additive ones. Further, for some pollutants, such as carcinogens, there are no established "safe" levels of exposure. Thus, while when risk assessments are carried out, they often show "acceptable risks," indicating no need for extraordinary controls, the ability to make such judgments in a meaningful way is seriously constrained.

- Should there be a fixed standard limit for all plants and each parameter, or should limits be determined case-by-case or by size? Fixed emissions rules, which may be stringent in light of then-existing technologies when first promulgated, cannot reflect evolving technologies or the variety of site-specific circumstances. Size-based rules tend to mandate less stringent standards for smaller plants. Thus, despite the demonstrated success of state-of-the-art technologies for such plants, size-based rules could tend to cause a proliferation of smaller, less well-controlled plants.

- Should federal and state emission limit statutes be periodically reevaluated so that more stringent requirements to permit new facilities could be considered as new technologies evolve?

Monitoring requirements are also critical. While the new federal regulations mandate continuous monitoring for oxides of nitrogen, sulfur dioxide, carbon monoxide, and opacity, they require at most only annual stack tests for particulates, dioxins and furans, and hydrogen chloride. Further, the required averaging time for the carbon monoxide tests ranges from 4 to 24 hours, and for the oxides of nitrogen and sulfur dioxide tests it is 24 hours, long enough periods to mask numerous and/or extreme short-term exceedences of the established limits. Since high carbon monoxide levels (more than 100 parts per million) indicate poor combustion, which is correlated with dioxin formation, masking carbon monoxide exceedences could mask high dioxin emission levels.

Worker training, essential to state-of-the-art plant operations, is not adequately covered in the new regulations, nor is there a plan to develop requirements for appropriate formal and practical educational programs.

Finally, regardless of the new national regulations, states have always had the right to impose more stringent standards. Thus, the potential for regulatory and informational confusion will continue, albeit to a lesser extent.

Nevertheless, several strategies could make it easier to compare plants operating under different regulations.

- National standards for expressing emissions measurements and permit conditions could be established even for pollutants that are not regulated nationally.
- States could be required to collect specific types of information and to make it available to the public in a single location within the state.
- A national clearinghouse could provide comparable facts, specifically including source separation and emissions figures, for each state and plant. Such information could be made public annually through one central location in each state as well as through a centralized national clearinghouse. While the EPA currently has a BACT/LAER Clearinghouse, it simply makes information about how control technology decisions have been made available to regulatory agencies, but does not provide the public with comprehensive plant-by-plant data about the environmental performance of individual incinerators.

Ash Regulations

Regulations for managing ash from garbage-burning plants have been slow in coming. Only recently have the issues of ash composition and toxicity been added to the list of concerns associated with incinerating municipal solid waste. Currently, ash is virtually unregulated on the federal level and regulations are only beginning to be proposed at the state level. Methods for assessing the toxicity of ash are in dispute; this absence of agreement on testing methodologies also precludes agreement on when ash fails a test and what disposal techniques should be used when it does. The diversity of existing regulations and practices reflects unresolved controversies over ash management.

Federal Regulations

At the federal level, confusion over ash regulations has resulted from conflicting signals about the classification of ash from garbage-burning plants. The Resource Conservation and Recovery Act (RCRA) mandates that materials exceeding the levels established by the Environmental Protection Agency's EP Tox test (discussed in Chapter 3) be considered hazardous and thus subject to certain handling and disposal regulations. However, this law has been read to exempt household waste and its products, such as incinerator ash, from consideration as legally defined hazardous waste, regardless of the toxic materials present in it. Recent court decisions have upheld this reading.[1] Thus, even though, according to a 1988 Environmental Defense Fund study, "virtually every sample of fly ash ever tested using the EP Toxicity test has exceeded the limits for lead or cadmium (usually both) defining hazardous waste,"[2] ash from the combustion of municipal solid waste in a waste-to-energy facility may be disposed of in an ordinary landfill, like any nonhazardous waste, except where state regulations prohibit this practice.

The Resource Conservation and Recovery Act is up for reauthorization in the 1991-1992 congressional session. The issue of establishing specific procedures for classifying and managing incinerator ash is likely to be debated during the reauthorization process.

State Regulations

With this essential absence of federal regulations, many states have adopted their own standards for ash management. These include different classifications of ash as well as ash disposal and testing regulations.

Six of the ten states INFORM studied have classified ash from garbage-burning plants separately from municipal solid waste. Maine and Oklahoma classify ash as "special waste," and Florida, New Hampshire, New York, and Oregon call it "ash residue." (They can do this, despite federal court decisions, because more stringent state regulations can supersede federal regulations.) In California, ash is classified as solid waste unless it fails the state's own toxicity tests.

Regulations for new ash disposal facilities in the ten states involve a variety of liner, leachate collection and treatment, and leak detection systems; all except Maryland and New York restrict ash disposal to monofills. However, while New York does not require monofill disposal, it mandates the use of double liners for ash that is codisposed with solid waste. Seven states have specific regulations for ash; the remaining three (Delaware, Maryland, and Mississippi) cover ash disposal under their general solid waste regulations. New York is unique in stipulating different disposal practices for fly ash and combined fly and bottom ash. Table 6-4 summarizes these disposal requirements.

Ash testing is required in all but two (Florida and Maryland) of the ten states, as Table 6-4 also shows. Seven of these states specify the EP Tox procedure which tests for eight metals: arsenic, barium, cadmium, chromium, lead, mercury, selenium, and silver. This is likely to change now that, as of early 1991, the EPA has selected the Toxic Characteristics Leaching Procedure (TCLP) test, rather than EP Tox, as its preferred ash testing procedure. (California uses a different procedure, the California Wet Extraction Procedure.) However, some states require testing only for those metals suspected of exceeding EP Tox levels, so that not all eight parameters are always tested. California and Maine also mandate testing for additional parameters (see Table 6-4). Despite these testing requirements, only California specifies different management procedures based on the results of the test.

Issues of Ash Regulation

Two key interrelated issues face regulators trying to establish standards for managing ash from garbage-burning plants: how to determine the toxicity of ash and how to classify it. Unless presorting to remove wastes containing toxic materials and pollutant precursors from the incinerator feedstock takes place, both issues will become increasingly critical as more ash is produced and as air emissions control devices become more effective at trapping toxic materials in fly ash.

Reliable methods for determining the toxicity of ash are important because ash composition changes constantly due to variations in waste composition, combustion efficiency, and emission control operations. However, as discussed in Chapter 3, there is no agreement about which, if any, of the existing testing procedures, including the newly preferred TCLP tests, provides acceptable information about ash toxicity. The EP Tox test, for example, although it was widely used, is a laboratory procedure, rather than a test under real ashfill conditions. The levels of pollutants found in leachate from ashfills did not always match those obtained in EP Tox lab tests. Further, it is not understood how heavy metals in leachate actually enter the environment and in what quantities.

Without agreement about reliable testing procedures, not only is classification difficult, but it is also impossible to determine the safety of the various ash reuse options. Use of these options would require specialized tests that could provide reasonable estimates for long-term deterioration in different environments, as well as for short-term impacts of demolition and disposal.

A fundamental issue involves whether it is feasible to repeatedly sample ash from a facility to determine how to handle and where to send each sample. Not only is this logistically difficult, but there is no agreement about how to determine an adequate sample size or sampling procedure. Instead, regulators could mandate ash handling, treatment, and disposal requirements based on current knowledge, rather than on the results of individual tests.

The debate over how to classify ash, as hazardous or nonhazardous waste, or as some intermediate special waste, is fueled significantly by the issue of costs. Materials legally classified as "hazardous" can only be disposed of in special "hazardous waste landfills," with truck manifesting, following the regulations established by the federal Resource Conservation and Recovery Act, with costs significantly higher than those for disposal of waste in an ordinary landfill. In New Jersey, for example, the cost of burying ash from the Warren County incinerator (not included in this study) in a hazardous waste landfill has been $250 per ton, almost four times the disposal cost for ash that is not legally classified as hazardous.[3]

If new federal or state regulations specify that ash can be considered hazardous if it exceeds limits for toxic content, separation of the more toxic fly ash from the bottom ash could potentially reduce the volume of ash requiring costly hazardous waste disposal procedures. If it is frequently and reliably found to be nontoxic, the bottom ash could potentially be reused, as discussed in Chapter 4, or disposed of in an ash monofill. Again, reliable and widely accepted testing techniques are required for these options to become a reality. Additionally, most existing waste-to-energy plants, including every facility in this study, combine the two ash streams, so in-plant modifications would be necessary to permit separate testing and disposal. A pilot study at an existing facility could shed light on the feasible approaches.

Table 6-4 Ash Regulations (1988)

	Ash Classification				
State	Solid waste	Special waste	Ash residue	Restricted to Monofill?	Number of Liners
California	■			Probably (undecided)	**
Delaware	■			■	1
Florida			■	■	1
Maine		■		■	2
Maryland	■			■	1
Mississippi	■			■‡	1
New Hampshire			■	■	2
New York***			■		2
Oklahoma		■		■	1
Oregon			■	■	1

* *All testing is for combined fly and bottom ash except Maine, which mandates separate testing, and New York, which mandates separate testing unless the ash streams are already combined.*

† *All states that require testing, except California, used the EP Tox test. (With EPA's new preference for the Toxic Characteristics Leaching Procedure test, this will begin to change.) California used the California Wet Extraction Procedure. California and Maine also specify additional parameters that must be tested (California: antimony, asbestos, beryllium, cobalt, copper, fluoride salts, hexavalent chromium, molybdenum, nickel, thallium, vanadium, and zinc; Maine: aluminum, beryllium, calcium, chloride, copper, hexavalent chromium, iron, magnesium, manganese, molybdenum, nickel, potassium, dioxins and furans, sulfur oxides, sodium, vanadium, zinc, pH, moisture percent, and carbon percent).*

Leachate Regulations		Groundwater Monitoring	Testing*,†	
Collection	Treatment	Required	Required?	Frequency per year
**	**	■	■	1-12×
■	■	■	■	4×
■	■	■		
■	■	■	■	4×
■	■	■		
■	■	■	■	4×
■	■	■	■	1-12×
■	■	■	■	2×
		■	■	4×
■		■	■	1-4×

** *Site-specific.*

‡ *And inert industrial waste.*

*** *Requirements are for combined ash codisposed with municipal solid waste. New York regulations specify a single liner if combined ash is disposed of in a monofill, and a monofill with a double liner for untreated fly ash, but they do not require ash to be separated from municipal solid waste.*

Notes

1 EDF vs. Wheelabrator Technologies, Inc, 3OERC1609 (SDNY November 21, 1989), and EDF vs. Chicago, 3OERC1624 (NDILL, November 29, 1989).

2 Environmental Defense Fund, *The Hazards of Ash and Fundamental Objectives of Ash Management*, 1988.

3 *The New York Times*, January 25, 1989, page B1.

APPENDIX A: PLANT PROFILES

ALBANY

Sheridan Avenue Refuse Derived Fuel Steam Plant
79 Sheridan Avenue
Albany, New York 12207

Plant Characteristics

Owner/operator
New York State Office of General Services

Vendor
New York State Office of General Services

Start-up date
February 1982

Type
Refuse-derived fuel

Type of fuel
Refuse-derived fuel, natural gas

Energy products
Steam

Customer
City of Albany (used to heat downtown Albany)

Energy rating
Plant did not respond

Design capacity
600 tons per day

Capacity being used
400 tons per day

Worker Training and Experience

Number of employees
Plant did not respond

Training
Plant did not respond

Experience

Chief facility operator
Plant did not respond

Shift supervisor
Plant did not respond

Control room operator
Plant did not respond

Service Area Characteristics

Area served
City of Albany

Population of service area
285,570

Amount of municipal solid waste generated in service area
348,600 tons per year
63% residential
37% commercial

Solid waste management in area
As of November 1988
59% landfilled
30% incinerated
7% recycled, 4% composted

Materials collected for recycling
Returnable beverage containers, tires, newspaper, high-grade paper, white goods

Planning

Criteria/method for sizing and design
1978 analysis based on area population and solid waste generation trends

Volume
Weight records

Composition
Sampling

Lab analysis
Plant did not respond

Siting

Nearest residence
1 block

Location
Existing steam-producing facility

Citizen involvement in planning
Public meetings and hearings

Citizen opposition
Environmental Planning Lobby, Shereton Hollow Neighborhood Association

Plant Design and Operations

Garbage storage
2 pits

Capacity (tons)
600

Capacity (days)
1

Screening of prohibited wastes

Technique
Visual

Responsible staff
Pickers and tipping floor person

Materials prohibited
Large metal objects, carpets, fencing cables

Percent rejected
5%

Penalty type
Suspension of permits

Penalty enforced by
Commissioner of Public Works

Penalty ever levied?
Plant did not respond

Furnace design

Loading technique
Continuous loading

Basic type
Single-chambered waterwall furnace with excess air

Modifications
Plant did not respond

Auxiliary burner
Yes

Operating temperature
2500°F

Automatic combustion controls
Yes

Emissions control equipment
Three-field electrostatic precipitator

Monitoring and Maintenance

Monitoring

Parameters monitored
Temperature, opacity, combustion efficiency, steam pressure

Monitors connected to alarms?
Plant did not respond

Parameters recorded
Plant did not respond

Time records kept
Plant did not respond

Frequency of reporting
Plant did not respond

Monitor calibration frequency
Plant did not respond

Maintenance frequency

Furnace
Plant did not respond

Boiler
Plant did not respond

Stoker and grate
Plant did not respond

Air pollution control equipment
Plant did not respond

Turbine
None

Other sections
Plant did not respond

Air Emissions

Date of test(s)
March 1987

Test(s) conducted by
New York State Department of Environmental Conservation

Emissions tested for
Particulates, total polychlorinated dibenzodioxins/polychlorinated dibenzofurans, hydrogen chloride, sulfur dioxide, oxides of nitrogen, arsenic, beryllium, cadmium, chromium, lead, manganese, mercury, nickel, vanadium, zinc

Particulates
0.020 grains per dry standard cubic foot at 7% O_2

Lead
1.28×10^{-3} grams per dry normal cubic meter at 7% O_2

Mercury
5.77×10^{-4} grams per dry normal cubic meter at 7% O_2

Dioxin/furan equivalents
18.8 nanograms per dry normal cubic meter at 7% O_2

Hydrogen chloride
464 parts per million at 7% O_2

Sulfur dioxide
224 parts per million at 7% O_2

Oxides of nitrogen
310 parts per million at 7% O_2

ASH

Ash amounts

Weight per day
110 tons per day

Volume per day
120 cubic yards

As % of original weight
28%

As % of original volume (estimate)
20-25%

Ash testing

Extraction Procedure Toxicity test, Toxic Characteristics Leaching Procedure

Materials tested for
Arsenic, barium, cadmium, chromium, lead, mercury, selenium, silver, combustibles, metals

Frequency of testing
June 1987 - July 1988

Ash handling and transportation

Ash handling
Combined (fly and bottom ash)

Ash covered in plant?
No

Ash covered while transported?
Yes

Mode of ash transportation
Covered, but not leakproof, dump truck

Ash treatment

Awaiting approval to process ash into aggregate substitute

Ash disposal

Albany City Landfill

Monofill or codisposal
Codisposal

Landfill liners
None

Leachate collection
None

Leachate treatment
None

Expected life of landfill
Until 1988; expansion was being planned

Length of ash disposal agreement
Plant did not respond

OTHER ENVIRONMENTAL IMPACTS

Truck traffic

Number of trucks per day

Delivering municipal solid waste
40

Removing ash
4-6

Pass through residential areas?
Yes

Water management

Amount used per day
Plant did not respond

Used for
Plant did not respond

Source
Plant did not respond

Disposal
Plant did not respond

Economic Factors

Costs

Capital construction (excluding financing)
$16 million (1981 dollars)

Operations per year (excluding ash management)
Plant did not respond

Revenues

Tipping fees (per ton)
Plant did not respond

Electricity (per kilowatt hour)
Not produced at plant

Steam (per 1000 pounds)
Plant did not respond

Scrap (per ton)
Not produced at plant

Control over revenues

Flow control?
Plant did not respond

Length of contract

Garbage
Plant did not respond

Electricity
Not produced at plant

Steam
Plant did not respond

AUBURN

Auburn Energy Recovery Facility
1 Goldthwaite Road
Auburn, Maine 04210

Plant Characteristics

Owner/operator
City of Auburn

Vendor
Consumat

Start-up date
April 1981

Shut-down date
February 15, 1990

Type
Mass burn

Type of fuel
Municipal solid waste

Energy products
Steam

Customer
Pioneer Plastic

Energy rating
50,000 pounds per hour

Design capacity
200 tons per day

Capacity being used
185 tons per day

Worker Training & Experience

Number of employees
31

Training
Formal training for all positions provided at Central Maine Vocational Technical Institute, Auburn, Maine

Experience

Chief facility operator
State-licensed boiler manager

Shift supervisor
State-licensed boiler engineer

Control room operator
Plant did not respond

Service Area Characteristics

Area served
Auburn, Maine

Population of service area
Approximately 65,000

Amount of municipal solid waste generated in service area
150-275 tons per day (seasonal variation)
60% residential
10% commercial
30% industrial

Solid waste management in area
30% landfilled
65% incinerated
5% recycled

Materials collected for recycling
Paper, glass

Planning

Criteria/method for sizing and design

Volume
Estimates from commercial waste haulers; waste records

Composition
Plant did not respond

Lab analysis
None

Siting

Nearest residence
Less than 1/4 mile

Location
Formerly a storage field adjacent to the steam customer

Citizen involvement in planning
Information through newspapers, radio, television

Citizen opposition
Local residents and watchdog organizations

Plant Design and Operations

Garbage storage
Tip floor

Capacity (tons)
400

Capacity (days)
2

Screening of prohibited wastes

Technique
Visual from tip floor

Responsible staff
Loader operators

Materials prohibited
Combustibles; oversized materials; liquid, chemical, infectious, pathological, and radioactive materials

Percent rejected
3%

Penalty type
None

Penalty enforced by
Department of Environmental Protection, Maine

Penalty ever levied?
No

Furnace design

Loading technique
Batch loading

Basic type
Modular furnace, dual-chambered, controlled air

Modifications
Plant did not respond

Auxiliary burner
Never operated

Operating temperature
1800°F

Automatic combustion controls
For secondary chamber temperature

Emissions control equipment
None; fabric filters were originally installed but were nonfunctional

Monitoring and Maintenance

Monitoring

Parameters monitored
Temperature, inlet and outlet temperatures for air pollution control devices, steam pressure

Monitors connected to alarms?
No

Parameters recorded
None

Time records kept
Not kept

Frequency of reporting
Not required

Monitor calibration frequency
No schedule

Maintenance frequency

Furnace
Monthly

Boiler
Weekly

Stoker and grate
Quarterly

Air pollution control equipment
None

Turbine
Plant did not respond

Other sections
Monthly

Air Emissions

Date of test(s)
October, March 1982

Test(s) conducted by
ETS, Inc.

Emissions tested for
Particulates

Particulates
0.08 grains per dry standard cubic foot at 12% CO_2

Ash

Ash amounts

Weight per day
120 tons (wet)

Volume per day
225 cubic yards

As % of original weight
Approximately 50% (wet)

As % of original volume (estimated)
10-12%

Ash testing

Materials tested for
Plant did not respond

Frequency of testing
Plant did not respond

Ash handling and transportation

Ash handling
Combined (fly and bottom ash)

Ash covered in plant?
No

Ash covered while transported?
Yes

Mode of ash transportation
Leakproof, covered trucks

Ash treatment
None

Ash disposal

Monofill or codisposal
Monofill

Landfill liners
Clay liner (one)

Leachate collection
Yes

Leachate treatment
Processed at sludge treatment plant

Expected life of landfill
10 weeks as of 9/23/88

Length of ash disposal agreement
Plant did not respond

Other Environmental Impacts

Truck traffic

Number of trucks per day

Delivering municipal solid waste
42-60

Removing ash
7

Pass through residential areas?
Yes

Water management

Amount used per day
1000 gallons

Used for
Quenching ash, cleaning plant

Source
City drinking water supply

Disposal
Sewer system without treatment

Economic Factors

Costs

Capital construction (excluding financing)
$3.98 million (1981 dollars)

Operations per year (excluding ash management)
$1.5 million

Revenues

Tipping fees (per ton)
$47.00 (members), $64.00 (non-members), $100.00 (special handling)

Electricity (per kilowatt hour)
Not produced at plant

Steam (per 1000 pounds)
Plant did not respond

Scrap (per ton)
Not produced at plant

Control over revenues

Flow control?
Yes

Length of contract

Garbage
20 years

Electricity
Not produced at plant

Steam
20 years

BALTIMORE

Baltimore Refuse Energy System Co.
1801 Annapolis Road
Baltimore, Maryland

Plant Characteristics

Owner/operator
Baltimore Refuse Energy System Company Ltd./Wheelabrator Baltimore Inc.

Vendor
Signal Environmental Systems, Inc.

Start-up date
May 1985

Type
Mass burn

Type of fuel
Municipal solid waste, commercial waste, natural gas for auxiliary fuel

Energy products
Steam, electricity

Customer
Steam to Baltimore Thermal Corp., electricity to Baltimore Gas & Electric Co.

Energy rating
55 megawatts

Design capacity
2250 tons per day

Capacity being used
2250 tons per day (as of 1988)

Worker Training and Experience

Number of employees
70 (6 operators per shift)

Training
Operator training required by state of Maryland Boiler Operator's License Bureau

Experience

Chief facility operator
Internal training program

Shift supervisor
Internal training program

Control room operator
Internal training program

Service Area Characteristics

Area served
Baltimore County, Annapolis

Population of service area
1,722,000 in 1985

Amount of municipal solid waste generated in service area
7000 tons per day (1988)
55% residential
30% commercial
15% industrial

Solid waste management in area
Plant did not respond for % landfilled; 26% incinerated; plant responded that no figures were available for % recycled

Materials collected for recycling
Newspapers, paper, aluminum, glass, corrugated

Planning

Criteria/method for sizing and design

Volume
Weight records and estimates of tonnages received at sites in the service area

Composition
None

Lab analysis
None

Siting

Nearest residence
0.5 miles

Location
Site of existing pyrolysis incinerator built on harbor

Citizen involvement in planning
None; site already authorized for waste-to-energy facility

Citizen opposition
None

Plant Design and Operations

Garbage storage
Pit

Capacity (tons)
8000

Capacity (days)
3.5

Screening of prohibited wastes

Technique
Visual by crane operator

Responsible staff
Crane operator

Materials prohibited
Industrial sludge, hazardous and chemical waste, construction and demolition debris, explosives

Percent rejected
Less than 1%

Penalty type
No fines; warnings issued and possible bans

Penalty enforced by
The plant

Penalty ever levied?
No

Furnace design

Loading technique
Continuous loading

Basic type
Single-chambered waterwall furnace with excess air

Modifications
Plant did not respond

Auxiliary burner
For start up and maintaining temperature

Operating temperature
1800° - 2500°F

Automatic combustion controls
For steam production, oxygen in flue gas, fuel burnout on grate

Emissions control equipment
Four-field electrostatic precipitator

Monitoring and Maintenance

Monitoring

Parameters monitored
Temperature, oxygen, opacity, inlet and outlet temperature for air pollution control devices, steam pressure, steam flow, water feed

Monitors connected to alarms?
Yes

Parameters recorded
Same as monitored

Time records kept
2 years

Frequency of reporting
Quarterly

Monitor calibration frequency
Continuous

Maintenance frequency

Furnace
Four times per year

Boiler
Four times per year

Stoker and grate
Plant did not respond

Air pollution control equipment
Quarterly

Turbine
Plant did not respond

Other sections
Plant did not respond

Air Emissions

Date of test(s)
January 1985

Test(s) conducted by
Entropy Environmentalists

Emissions tested for
Particulates, carbon monoxide, sulfur

dioxide, oxides of nitrogen, chlorides, fluorides

Note The 3 numbers for each pollutant refer to individual tests of each of 3 furnace/boiler units.

Particulates
0.0019, 0.0043, 0.0010 grains per dry standard cubic foot at 12% CO_2

Carbon monoxide
33.5, 11.4, 16.3 parts per million at 7% O_2

Sulfur dioxide
125, 107, 124 parts per million at 7% O_2

Oxides of nitrogen
221, 194, 194 parts per million at 7% O_2

Ash

Ash amounts

Weight per day
639 tons per day

Volume per day
675 cubic yards (estimate)

As % of original weight
25%

As % of original volume (estimated)
10%

Ash testing
Extraction Procedure Toxicity test

Materials tested for
Arsenic, barium, cadmium, chromium, lead, mercury, selenium, silver, chlorinated pesticides

Frequency of testing
Once since 1986

Ash handling and transportation

Ash handling
Combined (fly and bottom ash)

Ash covered in plant?
Yes

Ash covered while transported?
Yes

Mode of ash transportation
Covered truck

Ash treatment
Recovery of ferrous and nonferrous metals

Ash disposal
Baltimore City landfills

Monofill or codisposal
Codisposal

Landfill liners
Double liners

Leachate collection
Yes

Leachate treatment
Plant did not respond

Expected life of landfill
20 years

Length of ash disposal agreement
2003 (20 years from 1983)

Other Environmental Factors

Truck traffic

Number of trucks per day

Delivering municipal solid waste
400

Removing ash
35

Pass through residential areas?
No

Water management

Amount used per day
Plant did not respond

Used for
Quenching ash, cleaning

Source
City water supply

Disposal
Incremental sewer discharges, harbor

Economic Factors

Costs

Capital construction (excluding financing)
$170 million (1985 dollars)

Operations per year (excluding ash management)
Plant did not respond

Revenues

Tipping fees (per ton)
$33.28 (municipal), $34.68 (commercial)

Electricity (per kilowatt hour)
Plant did not respond

Steam (per 1000 pounds)
Not sold by plant

Scrap (per ton)
Not produced at plant

Control over revenues

Flow control?
Yes

Length of contract

Garbage
20 years

Electricity
25 years

Steam
Not sold

BIDDEFORD/SACO

Maine Energy Recovery Company
P.O. Box 401, 3 Lincoln Street
Biddeford, Maine

Plant Characteristics

Owner/operator
KTI Holdings, Inc./KTI Operations, Inc.

Vendor
General Electric

Start-up date
May 1987

Type
Refuse-derived fuel

Type of fuel
Municipal solid waste, wood chips, natural gas, oil

Energy products
Electricity

Customer
Central Maine Power Company

Energy rating
22 megawatts

Design capacity
607 tons per day

Capacity being used
607 tons per day

Worker Training and Experience

Number of employees
85

Training
Extensive experience required, college degree and licenses required, intensive on-job training and certification

Experience

Chief facility operator
Boilerhouse operator license

Shift supervisor
Boilerhouse operator license, plus extensive power plant experience

Control room operator
Same as shift supervisor

Service Area Characteristics

Area served
Southern Maine

Population of service area
250,000 (1980)

Amount of municipal solid waste generated in service area
190,000 tons per year
Plant did not respond for % residential and % commercial

Solid waste management in area
20% landfilled
80% incinerated
State mandated recycling program as of October 1988

Materials collected for recycling
Ferrous metals, aluminum, glass, batteries

Planning

Criteria/method for sizing and design

Volume
Volume study

Composition
None

Lab analysis
Refuse-derived fuel sample analyzed

Siting

Nearest residence
400 feet

Location
Near steam customers; formerly a 19th century textile mill

Citizen involvement in planning
Municipal planning committee

Citizen opposition
Yes, but no organized opposition prior to selection of project

Plant Design and Operations

Garbage storage
Tip floor

Capacity (tons)
3000

Capacity (days)
4

Screening of prohibited wastes

Technique
Visual inspection

Responsible staff
Tip floor personnel

Materials prohibited
Hazardous materials, medical waste, sludges, oils, etc.

Percent rejected
18-20% before refuse-derived fuel process, 0-1% after refuse-derived fuel process

Penalty type
Carter excluded from facility

Penalty enforced by
Maine Department of Environmental Protection

Penalty ever levied?
No (as of 9/88)

Furnace design

Loading technique
Batch loading

Basic type
Single-chambered, waterwall furnace with excess air

Modifications
None

Auxiliary burner
Natural gas, #2 oil

Operating temperature
1800° - 2000°F

Automatic combustion controls
No

Emissions control equipment
Dry acid scrubbers, fabric filters

Monitoring and Maintenance

Monitoring

Parameters monitored
Temperature, oxygen, carbon monoxide, opacity, hydrogen chloride, sulfur dioxide, carbon dioxide

Monitors connected to alarms?
Yes, and computer link to state Department of Environmental Protection

Parameters recorded
Same as monitored

Time records kept
7 years

Frequency of reporting
Continuous

Monitor calibration frequency
Every 6 months

Maintenance frequency

Furnace
Plant did not respond

Boiler
Annual

Stoker and grate
Periodic cleaning, annual maintenance

Air pollution control equipment
Weekly inspection, continuous maintenance

Turbine
Annual

Air Emissions

Date of test(s)
September, November, December 1987

Test(s) conducted by
Engineering Science (September), Entropy Environmentalists (November, particulates; December, dioxins)

Emissions tested for
Particulates, particles (less than 2 microns), carbon monoxide, 2,3,7,8-tetrachlorinated dibenzo-*p*-dioxin, total polychlorinated dibenzodioxins/polychlorinated dibenzofurans, hydrogen chloride, sulfur dioxide, oxides of nitrogen, total chlori-

nated hydrocarbons, volatile organic compounds, lead, mercury
Note Multiple numbers for certain pollutants refer to multiple individual test results.

Particulates
0.0079, 0.0078 grains per dry standard cubic foot at 12% CO_2

Lead
7.7×10^{-5} grams per normal cubic meter at 7% O_2

Mercury
4.48×10^{-5} grams per normal cubic meter at 7% O_2

Carbon monoxide
81, 81, 82, 87 parts per million at 12% CO_2

Dioxin/furan equivalents
0.712 nanograms per normal cubic meter at 12% CO_2

Hydrogen chloride
2, 5, 9, 22 parts per million at 12% CO_2

Sulfur dioxide
3, 3, 3, 11 parts per million at 12% CO_2

Oxides of nitrogen
202 parts per million at 12% CO_2

Ash

Ash amounts

Weight per day
75 tons per day

Volume per day
Plant did not respond

As % of original weight
10% (estimated)

As % of original volume (estimated)
Plant did not respond

Ash testing
Extraction Procedure Toxicity test

Materials tested for
Arsenic, barium, cadmium, chromium, lead, mercury, selenium, silver, all heavy metals, moisture, % solids

Frequency of testing
Quarterly

Ash handling and transportation

Ash handling
Combined (fly and bottom ash)

Ash covered in plant?
Yes

Ash covered while transported?
Yes

Mode of ash transportation
Covered, leakproof truck

Ash treatment
None

Ash disposal
Ash dump/Hampden, Maine

Monofill or codisposal
Monofill

Landfill liners
Plastic liner, plus packed clay

Leachate collection
Yes

Leachate treatment
On-site treatment

Expected life of landfill
1991 (14 months from August 1990)

Length of ash disposal agreement
Plant did not respond

Other Environmental Factors

Truck traffic

Number of trucks per day

Delivering municipal solid waste
85-100

Removing ash
4-5 (glass and grit removal 7)

Pass through residential areas?
No

Water management

Amount used per day
65,000 gallons per day

Used for
Non-contact cooling, boiler blowdown, condenser

Source
Saco River

Disposal
Cooling water to Saco River, boiler water to sewer systems

Economic Factors

Costs

Capital construction (excluding financing)
$67 million (1987 dollars)

Operations per year (excluding ash management)
Plant did not respond

Revenues

Tipping fees (per ton)
$4.00 (Biddeford), $8.00-$10.00 (others)

Electricity (per kilowatt hour)
Plant did not respond

Steam (per 1000 pounds)
None

Scrap (per ton)
Not separated from ash

Control over revenues

Flow control?
Yes

Length of contract

Garbage
20 years

Electricity
20 years

Steam
Not sold

CLAREMONT

New Hampshire/Vermont Solid Waste Project
Grissom Lane
Claremont, New Hampshire 03743

Plant Characteristics

Owner/operator
Signal Environmental Systems/Claremont Company, L.P. (now Wheelabrator Technologies)

Vendor
Wheelabrator Technologies

Start-up date
June 1987

Type
Mass burn

Type of fuel
Municipal solid waste, light industrial waste, natural gas used for start up

Energy products
Electricity

Customer
Central Vermont Public Service Co.

Energy rating
4.5 megawatts

Design capacity
200 tons per day

Capacity being used
171 tons per day

Worker Training and Experience

Number of employees
12

Training
Formal training prior to starting work by in-house personnel and equipment suppliers

Experience
Licensed boiler operators, or licensed by state for hazardous waste identification. Plant did not respond specifically for positions of chief facility operator, shift supervisor, or control room operator.

Service Area Characteristics

Area served
New Hampshire/Vermont

Population of service area
72,800

Amount of municipal solid waste generated in service area
Plant did not provide information on total tonnage of waste
65% residential
35% commercial

Solid waste management in area
As of October 1988
20% landfilled
75% incinerated
5% recycled

Materials collected for recycling
Newspapers, bottles, aluminum cans, batteries

Planning

Criteria/method for sizing and design

Volume
Weight surveys of waste in the service area with other data from neighboring areas, and projections of population size

Composition
Field analysis

Lab analysis
None

Siting

Nearest residence
1/2 - 3/4 mile

Location
New

Citizen involvement in planning
Advisory committee participation in the planning stage

Citizen opposition
After plant already operating

Plant Design and Operations

Garbage storage
Tipping floor

Capacity (tons)
600

Capacity (days)
3

Screening of prohibited wastes

Technique
Visual observation of incoming waste

Responsible staff
Front-end loader operators

Materials prohibited
Hazardous wastes

Percent rejected
Less than 1%

Penalty type
Carters threatened with exclusion from facility

Penalty enforced by
Plant did not respond

Penalty ever levied?
No

Furnace design

Loading technique
Front-end loader, continuous loading

Basic type
Single-chambered furnace with excess air

Modifications
None

Auxiliary burner
Yes

Operating temperature
1800°F

Automatic combustion controls
For steam production and oxygen in flue gas

Emissions control equipment
Fabric filter, dry scrubber

Monitoring and Maintenance

Monitoring

Parameters monitored
Temperature, oxygen, carbon monoxide, opacity, steam pressure

Monitors connected to alarms?
Yes

Parameters recorded
Same as monitored

Time records kept
At least 2 years

Frequency of reporting
Quarterly reports

Monitor calibration frequency
Continuous

Maintenance frequency

Furnace
4 times per year

Boiler
4 times per year

Stoker and grate
4 times per year

Air pollution control equipment
4 times per year

Turbine
3-5 years

Other sections
Plant did not respond

Air Emissions

Date of test(s)
May, October 1987

Test(s) conducted by
Entropy Environmentalists

Emissions tested for
Particulates, carbon monoxide, total polychlorinated dibenzodioxins /polychlorinated dibenzofurans, hydrogen chloride, sulfur dioxide, oxides of nitrogen
Note The 2 numbers for each pollutant refer to individual tests of each of 2 furnace/boiler units.

Particulates
0.0111, 0.00427 grains per dry standard cubic foot at 12% CO_2

Carbon monoxide
49.8, 47.6 parts per million at 7% O_2

Hydrogen chloride
104, 36.5 parts per million at 7% O_2

Sulfur dioxide
230, 59.9 parts per million at 7% O_2

Oxides of nitrogen
249, 210 parts per million at 7% O_2

Ash

Ash amounts

Weight per day
61 tons per day

Volume per day
61 cubic yards

As % of original weight
35.7%

As % of original volume (estimated)
12%

Ash testing
Extraction Procedure Toxicity test

Materials tested for
Arsenic, barium, cadmium, chromium, lead, mercury, selenium, silver

Frequency of testing
Monthly

Ash handling and transportation

Ash handling
Combined (fly and bottom ash)

Ash covered in plant?
Yes

Ash covered while transported?
Yes

Mode of ash transportation
Leakproof, covered containers

Ash treatment
Fly ash is treated with a lead immobilization additive (phosphoric acid)

Ash disposal
BFI Rockingham Landfill, Springfield, VT

Monofill or codisposal
Monofill (separate cell in municipal solid waste landfill)

Landfill liners
Double liner

Leachate collection
Yes

Leachate treatment
To wastewater treatment plant

Expected life of landfill
2007 (20 years from 1987); built in phases with 20-year complete life expectancy; first phase, 3- to 5-year life

Length of ash disposal agreement
20 years

Other Environmental Factors

Truck traffic

Number of trucks per day

Delivering municipal solid waste
17

Removing ash
3

Pass through residential areas?
Plant did not respond

Water management

Amount used per day
Plant did not respond

Used for
Boiler and cooling tower

Source
City of Claremont

Disposal
No disposal; zero-discharge plant

Economic Factors

Costs

Capital construction (excluding financing)
$17.9 million (1987 dollars)

Operations per year (excluding ash management)
Plant responded that this was "private information"

Revenues

Tipping fees (per ton)
Plant did not respond

Electricity (per kilowatt hour)
$0.09

Steam (per 1000 pounds)
None

Scrap (per ton)
Not separated from ash

Control over revenues

Flow control?
Yes

Length of contract

Garbage
20 years

Electricity
20 years

Steam
Not sold

COMMERCE

Commerce Refuse to Energy Facility
5926 Sheila Street
Commerce, California

Plant Characteristics

Owner/operator
Commerce Refuse to Energy Authority (CREA) /County Sanitation Districts of Los Angeles County

Vendor
Facility was designed by HDR; Foster Wheeler provided boiler and air pollution control equipment

Start-up date
May 1987

Type
Mass burn

Type of fuel
Municipal solid waste (95% commercial, 5% residential), natural gas in auxiliary burners

Energy products
Electricity

Customer
Southern California Edison

Energy rating
11.44 megawatts

Design capacity
330 tons per day

Capacity being used
Plant operating at full capacity

Worker Training and Experience

Number of employees
32

Training
Classroom training and on-the-job training for shift supervisor, control room operator, crane operator, tipping floor operator, ash handler

Experience

Chief facility operator
Boiler operation experience

Shift supervisor
Boiler operation experience

Control room operator
Boiler operation experience

Service Area Characteristics

Area served
City of Commerce

Population of service area
11,800 in 1984

Amount of municipal solid waste generated in service area
404 tons per day
5% residential
95% commercial

Solid waste management in area
As of October 1988
98% incinerated
2% recycled

Materials collected for recycling
Steel

Planning

Criteria/method for sizing and design

Volume
Waste sampling; population, employment, and industry data

Composition
Waste sampling

Lab analysis
Prior to design

Siting

Nearest residence
300 yards

Location
Former vacant lot

Citizen involvement in planning
Industrial Advisory Committee

Citizen opposition
None at the time of planning

Plant Design and Operations

Garbage storage
Pit

Capacity (tons)
1200

Capacity (days)
4

Screening of prohibited wastes

Technique
Visual; radioactive meter; sampling

Responsible staff
Weigh master, plant personnel, bulldozer and crane operator

Materials prohibited
Lead acid batteries, oversized objects, and ferrous metals

Percent rejected
1.6%

Penalty type
$25,000 and loss of carter's license

Penalty enforced by
CREA

Penalty ever levied?
No

Furnace design

Loading technique
Continuous loading

Basic type
Single-chambered waterwall furnace with excess air

Modifications
No major modifications

Auxiliary burner
Yes

Operating temperature
1800°F

Automatic combustion controls
Yes

Emissions control equipment
Fabric filter, dry scrubber, Thermal $DeNo_x$

Monitoring and Maintenance

Monitoring

Parameters monitored
Temperature, oxygen, carbon monoxide, sulfur oxides, oxides of nitrogen, steam pressure

Monitors connected to alarms?
No

Parameters recorded
Same as monitored

Time records kept
Plant did not respond

Frequency of reporting
Every month

Monitor calibration frequency
Once a day

Maintenance frequency

Furnace
Plant did not respond

Boiler
Plant did not respond

Stoker and grate
Plant did not respond

Air pollution control equipment
Plant did not respond

Turbine
Plant did not respond

Other sections
Plant did not respond

Air Emissions

Date of test(s)
May-June 1987, September 1987

Test conducted by
Energy Systems Associates

Emissions tested for
Particulates, carbon monoxide, 2,3,7,8-tetrachlorinated dibenzo-*p*-dioxin (Eadon), total polychlorinated dibenzodioxins/polychlorinated dibenzofurans, sulfur dioxide, oxides of nitrogen, hydrogen chloride, hydrogen fluoride, hydrocarbons, total chlorinated

hydrocarbons, chlorobenzene, chlorophenol, antimony, arsenic, beryllium, cadmium, chromium, copper, lead, mercury, nickel, selenium, thallium, zinc

Particulates
0.0043 at 12% CO_2

Lead
Less than 0.0042×10^{-3} grams per normal cubic meter at 12% CO_2

Mercury
Less than 0.58×10^{-3} grams per normal cubic meter at 12% CO_2

Carbon monoxide
16 parts per million at 7% O_2

Dioxin/furan equivalents
0.027 nanograms per dry normal cubic meter at 7% O_2

Hydrogen chloride
8.9 parts per million at 7% O_2

Sulfur oxides
1.3 parts per million at 7% O_2

Oxides of nitrogen
90.2 parts per million at 7% O_2

Ash

Ash amounts

Weight per day
100 tons

Volume per day
Plant did not respond

As % of original weight
25%

As % of original volume (estimated)
10%

Ash testing
Extraction Procedure Toxicity test

Materials tested for
Arsenic, barium, cadmium, chromium, lead, mercury, selenium, silver

Frequency of testing
8 times between 9/87 and 8/88

Ash handling and transportation

Ash handling
Combined (fly and bottom ash)

Ash covered in plant?
No

Ash covered while transported?
Yes

Mode of ash transportation
Covered trucks

Ash treatment
Plant did not respond

Ash disposal
Puente Hills Landfill

Monofill or codisposal
Codisposal

Landfill liners
Not required

Leachate collection
Underground dams and monitoring wells

Leachate treatment
Chlorinated and then disposed of in sewer system

Expected life of landfill
1993 (5 years from 1988); 2013 if expanded

Length of ash disposal agreement
5 years

Other Environmental Impacts

Truck traffic

Number of trucks per day

Delivering municipal solid waste
66

Removing ash
8

Pass through residential areas?
Yes

Water management

Amount used per day
300,000 gallons per day

Used for
Cooling tower

Source
Municipal

Disposal
Water discharge to water treatment facility

ECONOMIC FACTORS

Costs

Capital construction (excluding financing)
$35,000,000 (1987 dollars)

Operations per year (excluding ash management)
$5,000,000

Revenues

Tipping fees (per ton)
$18.00

Electricity (per kilowatt hour)
$0.08

Steam (per 1000 pounds)
None

Scrap (per ton)
Not separated from ash

Control over revenues

Flow control?
Yes

Length of contract

Garbage
None

Electricity
30 years

Steam
Not sold

DADE COUNTY

Dade County Resource Recovery Facility
6990 NW 97
Miami, Florida 33178

PLANT CHARACTERISTICS

Owner/operator
Metropolitan Dade Co./Montenay Power Corporation

Vendor
Zurn Boilers, Heil Frontend, BBC Turbine Generators

Start-up date
June 1982 (original facility)
Reconstruction completed in 1990

Type
Refuse-derived fuel

Type of fuel
Originally wet refuse-derived fuel, now only dry refuse-derived fuel

Energy products
Electricity

Customer
Florida Power and Light

Energy rating
76 megawatts

Design capacity
3000 tons per day

Capacity being used
1700 tons per day during reconstruction
(2600 tons per day after reconstruction)

WORKER TRAINING AND EXPERIENCE

Number of employees
210

Training
On the job training by in-house personnel

Experience

Chief facility operator
Some engineering experience

Shift supervisor
Some engineering experience, on-the-job training

Control room operator
Some engineering experience, on-the-job training

SERVICE AREA CHARACTERISTICS

Area served
Dade County

Population of service area
1,982,000 (1990)

Amount of municipal solid waste generated in service area
75% residential
25% commercial

Solid waste management in area
As of 1990; little recycling at time of plant visit
52% landfilled
26% incinerated
22% recycled

Materials collected for recycling
Yard waste, newspapers, office paper, metals, aluminum, glass, plastic bottles

PLANNING

Criteria/method for sizing and design

Volume
Weight records

Composition
None

Lab analysis
None

Siting

Nearest residence
1 mile

Location
Site of an existing dump (Superfund site)

Citizen involvement in planning
None

Citizen opposition
After plant went into operation

Plant Design and Operations

Garbage storage
Tipping floor before reconstruction (two pits after reconstruction)

Capacity (tons)
3000 before reconstruction (6000 after reconstruction)

Capacity (days)
1 before reconstruction (2 after reconstruction)

Screening of prohibited wastes

Technique
Observation during dumping

Responsible staff
Waste management officials

Materials prohibited
Hazardous, demolition, infectious materials

Percent rejected
2% (40% rejected during RDF process)

Penalty type
Fines

Penalty enforced by
Waste management officials

Penalty ever levied?
No

Furnace design

Loading technique
Continuous loading

Basic type
Single-pass, waterwall furnace with a traveling grate and excess air

Modifications
After reconstruction new boilers, larger furnace area, new stoking equipment, larger electrostatic precipitators

Auxiliary burner
Gas

Operating temperature
1800°F

Automatic combustion controls
Yes

Emissions control equipment
Cyclone, three-field electrostatic precipitator

Monitoring and Maintenance

Monitoring

Parameters monitored
Temperature, opacity (before reconstruction)
Oxygen, carbon monoxide, carbon dioxide (after reconstruction)

Monitors connected to alarms?
No

Parameters recorded
Opacity

Time records kept
24 hours

Frequency of reporting
4 times per year

Monitor calibration frequency
Every 8 hours

Maintenance frequency

Furnace
Once a year, or sooner as inspections warrant

Boiler
Once a year, or sooner as inspections warrant

Stoker and grate
Every 6 months

Air pollution control equipment
Once a year

Turbine
Once every 2 years, 5 years for major overhaul

Other sections
Cranes weekly; processing equipment monthly

Air Emissions

Date of test(s)
April 1987, January 1988

Test(s) conducted by
South Environmental Services, Inc.

Emissions tested for
Particulates

Particulates
0.0258 grains per dry standard cubic foot at 12% CO_2

Ash

Ash amounts

Weight per day
308 tons per day

Volume per day
2933 cubic yards

As % of original weight
20%

As % of original volume (estimated)
5%

Ash testing
Extraction Procedure Toxicity test

Materials tested for
Arsenic, barium, cadmium, chromium, lead, mercury, selenium, silver, organics, moisture

Frequency of testing
Quarterly

Ash handling and transportation

Ash handling
Combined (fly and bottom ash)

Ash covered in plant?
Closed conveyer, closed storage building

Ash covered while transported?
No, transported only on-site

Mode of ash transportation
Trucks

Ash treatment
None

Ash disposal
Ash landfill on site

Monofill or codisposal
Monofill

Landfill liners
One

Leachate collection
Yes

Leachate treatment
Sewer system to treatment plant off-site

Expected life of landfill
2068 (80 years from 1988)

Length of ash disposal agreement
Own ashfill; no contract

Other Environmental Factors

Truck traffic

Number of trucks per day

Delivering municipal solid waste
Total 500; 10-20 on site at one time

Removing ash
2-3 at any one time

Pass through residential areas?
No; the plant is located near major highways

Water management

Amount used per day
Not recorded

Used for
Quenching ash

Source
Wells, city drinking water

Disposal
Sewer system

Economic Factors

Costs

Capital construction (excluding financing)
$165 million (1982 dollars), $65 million reconstruction

Operations per year (excluding ash management)
$22 million

Revenues

Tipping fees (per ton)
$22.00

Electricity (per kilowatt hour)
$0.02

Steam (per 1000 pounds)
Not sold

Scrap (per ton)
$40-$45

Control over revenues

Flow control?
Yes

Length of contract

Garbage
15 years

Electricity
None

Steam
Not sold

LAKELAND

The McIntosh Power Plant
3030 E. Lake Parker Drive
Lakeland, Florida 33805

Plant Characteristics

Owner/operator
City of Lakeland/City of Lakeland, 60%; Orlando Utilities, 40%

Vendor
Linder Machine, Inc. (Designer: Horner and Shiftner)

Start-up date
1983 (for refuse-derived fuel)

Type
90% high-sulfur pulverized coal, 10% refuse-derived fuel

Type of fuel
Coal, refuse-derived fuel (residential, commercial)

Energy products
Electricity

Customer
City of Lakeland, Orlando

Energy rating
364 megawatts

Design capacity
500 tons per day of refuse-derived fuel (originally 250-300 tons per day)

Capacity being used
78% of RDF capacity in 1987

Worker Training and Experience

Number of employees
36 (in RDF section)

Training
On-the-job training

Experience

Chief facility operator
Plant did not respond

Shift supervisor
Plant did not respond

Control room operator
Plant did not respond

Service Area Characteristics

Area served
City of Lakeland

Population of service area
100,000

Amount of municipal solid waste generated in service area
50,000 tons per year
49% residential
40% commercial
11% other

Solid waste management in area
42% landfilled
58% incinerated
No recycling as of 9/88

Materials collected for recycling
Plant did not respond

Planning

Criteria/method for sizing and design

Volume
Volume study taken from data collected from county landfill

Composition
Plant did not respond

Lab analysis
Plant did not respond

Siting

Nearest residence
2 miles

Location
Plant did not respond

Citizen involvement in planning
Public meetings

Citizen opposition
None

Plant Design and Operations

Garbage storage
Floor

Capacity (tons)
350

Capacity (days)
Less than a day

Screening of prohibited wastes

Technique
Waste picked over by hand on tipping floor

Responsible staff
Workers on tipping floor

Materials prohibited
Explosive materials, tires, mattresses, and "white goods" (large household appliances)

Percent rejected
10%

Penalty type
None

Penalty enforced by
Not applicable

Penalty ever levied?
Not applicable

Furnace design

Loading technique
Continuous loading

Basic type
Single-chambered waterwall furnace with excess air

Modifications
None

Auxiliary burner
No

Operating temperature
2400°F

Automatic combustion controls
No

Emissions control equipment
Five-field electrostatic precipitators, wet scrubbers

Monitoring and Maintenance

Monitoring

Parameters monitored
Temperature, oxygen, opacity, sulfur oxides, inlet and outlet temperature for air pollution control devices, steam pressure

Monitors connected to alarms?
For sulfur dioxide and opacity

Parameters recorded
Same as monitored

Time records kept
1 month on site; 5 years off site

Frequency of reporting
Quarterly

Monitor calibration frequency
Daily

Maintenance frequency

Furnace
2 times per year

Boiler
2 times per year

Stoker and grate
2 times per year

Air pollution control equipment
2 times per year

Turbine
2 times per year

Other sections
2 times per year

Air Emissions

Date of test(s)
June 1988

Test(s) conducted by
Environmental Science and Engineering, Inc.

Emissions tested for
Particulates, sulfur dioxide, oxides of nitrogen

Particulates
0.019 grains per dry standard cubic foot at 7% O_2

Sulfur dioxide
179.2 parts per million at 7% O_2

Oxides of nitrogen
184 parts per million at 7% O_2

Ash

Not recorded; ash from refuse-derived fuel is a very small percent of total ash from coal

Ash amounts

Weight per day
Not recorded

Volume per day
Not recorded

As % of original weight
Not recorded

As % of original volume (estimated)
Not recorded

Ash testing

No

Materials tested for
None

Frequency of testing
Never

Ash handling and transportation

Ash handling
Combined (fly and bottom ash)

Ash covered in plant?
No

Ash covered while transported?
No trucks are used

Mode of ash transportation
Conveyor system

Ash treatment

Mixed with scrubber sludge, quicklime

Ash disposal

Monofill or codisposal
Monofill

Landfill liners
None

Leachate collection
Runoff collection system only (ash cementation so leachate not formed)

Leachate treatment
Water recirculated

Expected life of landfill
2013 (30 years from 1983)

Length of ash disposal agreement
No agreement

Other Environmental Factors

Truck traffic

Number of trucks per day

Delivering municipal solid waste
40

Removing ash
None

Pass through residential areas?
No

Water management

Amount used per day
Plant did not respond

Used for
Plant did not respond

Source
Storage ponds; recirculated

Disposal
Water is treated before returning to sewage treatment plant

Economic Factors

Costs

Capital construction (excluding financing)
$5 million for refuse-derived fuel section (1983 dollars)

Operations per year (excluding ash management)
$500,000 for refuse-derived fuel section

Revenues

Tipping fees (per ton)
$12.00 (city), $16.25 (noncity)

Electricity (per kilowatt hour)
$0.07

Steam (per 1000 pounds)
Not sold

Scrap (per ton)
Not produced at plant

Control over revenues

Flow control?
No

Length of contract

Garbage
None

Electricity
None

Steam
Not sold

MARION COUNTY

Marion County Solid Waste to Energy Facility
4850 Brookdale Road, NE, P.O. Box 9126
Brooks, Oregon 97305

Plant Characteristics

Owner/operator
Ogden Martin Systems of Marion, Inc.

Vendor
Ogden Martin Systems of Marion, Inc.

Start-up date
May 1986

Type
Mass burn

Type of fuel
Municipal solid waste (residential and commercial waste), natural gas, hospital waste

Energy products
Electricity

Customer
Portland General Electric Co.

Energy rating
13.1 megawatts

Design capacity
550 tons per day at 4500 British thermal units per pound

Capacity being used
510 tons per day

Worker Training and Experience

Number of employees
36

Training
On-the-job and classroom training

Experience

Chief facility operator
Past experience

Shift supervisor
2 years in plant operation

Control room operator
2 years in plant operation

Service Area Characteristics

Area served
Marion County

Population of service area
210,000 (1988)

Amount of municipal solid waste generated in service area
185,000 tons per year (approximately 160,000 tons per year from Marion County and 25,000 tons per year from out-of-county)
Plant did not respond for % residential or % commercial

Solid waste management in area
0.5% landfilled
77% incinerated
22.5% recycled

Materials collected for recycling
Paper, cardboard, aluminum, glass, ferrous metal, some plastics, used motor oil, ferrous metal recovered from ash streams, returnable beverage containers

Planning

Criteria/method for sizing and design

Volume
Scale records at Brown's Island and Woodburn landfills

Composition
Waste composition study, Marion County Public Works Department, 1981

Lab analysis
None

Siting

Nearest residence
1/4 mile

Location
New; formerly farm land (zoned for industrial use), deserted housing

Citizen involvement in planning
Public education; Solid Waste Advisory Committee (a citizens' advisory board) set up to recommend solid waste management priorities to county

Citizen opposition
1983 initiative petition which was defeated

Plant Design and Operations

Garbage storage
Pit

Capacity (tons)
2000

Capacity (days)
4

Screening of prohibited wastes

Technique
Visual

Responsible staff
Equipment operator, crane operator

Materials prohibited
Construction and demolition debris, hazardous and radioactive wastes, explosives, sewage sludge, bulky materials

Percent rejected
0.5%

Penalty type
Repeat offenders risk possibility of losing trucking license

Penalty enforced by
Department of Environmental Quality (Oregon)

Penalty ever levied?
No (as of July, 1990)

Furnace design

Loading technique
Continuous loading by cranes

Basic type
Single-chamber waterwall furnace with excess air

Modifications
None

Auxiliary burner
Yes

Operating temperature
1800°F minimum

Automatic combustion controls
Yes

Emissions control equipment
Dry scrubber, fabric filter

Monitoring and Maintenance

Monitoring

Parameters monitored
Temperature, oxygen, opacity, inlet and outlet temperatures for air pollution control devices, steam pressure, steam flow (monitors for sulfur dioxide and oxides of nitrogen added in late 1989)

Monitors connected to alarms?
For opacity, temperature, oxygen, and sulfur dioxide

Parameters recorded
Same as monitored

Time records kept
3 years

Frequency of reporting
Monthly

Monitor calibration frequency
Opacity monitors calibrated every 12 hours; oxygen monitors calibrated every 24 hours; sulfur dioxide, oxides of nitrogen, and temperature monitors calibrated every 24 hours

Maintenance frequency

Furnace
At 1/2 year intervals

Boiler
At 1/2 year intervals

Stoker and grate
At 1/2 year intervals

Air pollution control equipment
At 1/2 year intervals

Turbine
Every 3-5 years.

Other sections
As required, or 1/2 year intervals

AIR EMISSIONS

Date of test(s)
September, October 1986

Test(s) conducted by
Ogden Martin and US EPA test team

Emissions tested for
Particulates, carbon monoxide, total polychlorinated dibenzodioxins/polychlorinated dibenzofurans, 2,3,7,8-tetrachlorinated dibenzo-*p*-dioxin, hydrogen chloride, sulfur dioxide, oxides of nitrogen, fluorides, hydrogen fluoride, volatile organic compounds, beryllium, lead, mercury
Note When 2 numbers are given for a pollutant, they refer to individual tests of the 2 furnace/boiler units.

Particulates
0.011, 0.003 grains per dry standard cubic foot at 12% CO_2

Lead
2.50×10^{-5} grams per normal cubic meter at 12% CO_2

Mercury
0.24×10^{-3}, 0.32×10^{-3} grams per normal cubic meter at 12% CO_2

Carbon monoxide
16, 21 parts per million dry volume at 12% CO_2

Dioxin/furan equivalents
0.155 nanograms per normal cubic meter at 12% CO_2

Hydrogen chloride
3.94, 20.0 parts per million dry volume at 12% CO_2

Sulfur dioxide
38, 45 parts per million dry volume at 12% CO_2

Oxides of nitrogen
306, 283 parts per million dry volume at 12% CO_2

ASH

Ash amounts

Weight per day
120-125 tons per day

Volume per day
Not measured

As % of original weight
22%

As % of original volume (estimated)
5-10%

Ash testing
Extraction Procedure Toxicity test

Materials tested for
Arsenic, barium, cadmium, chromium, lead, mercury, selenium, silver

Frequency of testing
Plant responded that no further testing was required

Ash handling and transportation

Ash handling
Combined (fly and bottom ash)

Ash covered in plant?
Yes

Ash covered while transported?
Yes; covered on trailers

Mode of ash transportation
Leakproof, covered trucks

Ash treatment
None

Ash disposal

Monofill or codisposal
Monofill

Landfill liners
2

Leachate collection
Yes

Leachate treatment
Leachate diluted with fresh water to lower conductivity and then used as a spray irrigant

Expected life of landfill
1995 (5 years from July 1990); monofill 1998 (8 years from July, 1990)

Length of ash disposal agreement
Plant did not respond

Other Environmental Factors

Truck traffic

Number of trucks per day

Delivering municipal solid waste
80-120 weekdays/10-30 weekends

Removing ash
6-10 (5 days a week)

Pass through residential areas?
No; plant is off the main highway

Water management

Amount used per day
No measurement

Used for
Quenching ash, boiler blowdown, steam makeup, sanitizing

Source
Wells on site

Disposal
Into the Willamette River under National Pollution Discharge Elimination System (NPDES) permit

Economic Factors

Costs

Capital construction (excluding financing)
$47.5 million (1986 dollars)

Operations per year (excluding ash management)
$3,250,000 plus escalation and pass-through costs (base year for costs not clearly stated by plant)

Revenues

Tipping fees (per ton)
$26.00

Electricity (per kilowatt hour)
$0.06

Steam (per 1000 pounds)
Not sold

Scrap (per ton)
$34.00

Control over revenues

Flow control?
Yes

Length of contract

Garbage
20 years

Electricity
20 years

Steam
Not sold

OSWEGO

Oswego County Energy Recovery Facility
RR #2 Box 184 A
Fulton, New York 13069

Plant Characteristics

Owner/operator
Oswego County

Vendor
Consumat

Start-up date
February 1986

Type
Mass burn

Type of fuel
Municipal solid waste, light industrial waste, natural gas for auxiliary burners

Energy products
Steam, electricity

Customer
Steam to Armstrong World Industries, electricity to Niagara Mohawk Power Corporation

Energy rating
4 megawatts

Design capacity
200 tons per day

Capacity being used
190 tons per day

Worker Training and Experience

Number of employees
30

Training
Six weeks classroom training by manufacturer, 2 months on-the-job training

Experience

Chief facility operator
Engineer; has worked on several garbage-burning plant designs, took American Society of Mechanical Engineers course to certify chief plant operators in New York State

Shift supervisor
Background in steam plant or related field

Control room operator
On-the-job training

Service Area Characteristics

Area served
Oswego County

Population of service area
113,000

Amount of municipal solid waste generated in service area
197,000 tons per year
30% residential
70% commercial

Solid waste management in area
As of October 1988
65% landfilled
34% incinerated
1% recycled

Materials collected for recycling
Metals, newsprint (removed because of difficulty finding markets)

Planning

Criteria/method for sizing and design

Volume
Yes

Composition
Yes

Lab analysis
No

Siting

Nearest residence
1/2 mile

Location
Former vacant lot

Citizen involvement in planning
Public meetings

Citizen opposition
Complaints due to odor and dust after plant operation began

Plant Design and Operations

Garbage storage
Floor

Capacity (tons)
1000

Capacity (days)
5

Screening of prohibited wastes

Technique
Visual

Responsible staff
Front-end loader operator

Materials prohibited
Lead-acid batteries, oversized objects

Percent rejected
2-3%

Penalty type
Rejection of entire load

Penalty enforced by
Plant personnel

Penalty ever levied?
Once a month

Furnace design

Loading technique
Batch loading

Basic type
Modular dual-chambered, controlled air

Modifications
None

Auxiliary burner
Yes

Operating temperature
1800°F

Automatic combustion controls
For combustion temperature

Emissions control equipment
Two-field electrostatic precipitator

Monitoring and Maintenance

Monitoring

Parameters monitored
Temperature, oxygen, carbon monoxide, inlet and outlet temperature for air pollution control devices, steam pressure, steam flow, air pollution control voltage and amperage

Monitors connected to alarms?
Yes

Parameters recorded
Same as monitored

Time records kept
Telemetered to central computer off-site; plant did not provide information on how long computer records kept

Frequency of reporting
Not sent to regulatory agency

Monitor calibration frequency
Monthly

Maintenance frequency

Furnace
4 week cycle

Boiler
2 week cycle

Stoker and grate
4 week cycle

Air pollution control equipment
4 week cycle

Turbine
Yearly

Other sections
Normal preventive maintenance as for any other mechanical devices

Air Emissions

Date of test(s)
September 1986

Test(s) conducted by
New York State Department of Environmental Conservation

Emissions tested for
Particulates, carbon monoxide, 2,3,7,8-tetrachlorinated dibenzo-*p*-dioxin, hydrogen chloride, sulfur dioxide, oxides of

nitrogen, arsenic, beryllium, cadmium, chromium, lead, manganese, mercury, nickel, vanadium, zinc
Note Only 1 of 4 stacks tested

Particulates
0.013 grains per dry standard cubic foot at 12% CO_2

Lead
8.48×10^{-4} grams per dry normal cubic meter at 7% O_2

Mercury
6.98×10^{-4} grams per dry normal cubic meter at 7% O_2

Carbon monoxide
Less than or equal to 20 parts per million at 7% O_2

Dioxin/furan equivalents
11.23 nanograms per dry normal cubic meter at 7% O_2

Hydrogen chloride
552 parts per million at 7% O_2

Sulfur dioxide
389 parts per million at 7% O_2

Oxides of nitrogen
197 parts per million at 7% O_2

ASH

Ash amounts

Weight per day
75 tons per day

Volume per day
140 cubic yards

As % of original weight
40%

As % of original volume (estimated)
10%

Ash testing
Extraction Procedure Toxicity test, Toxic Characteristics Leaching Procedure

Materials tested for
Arsenic, barium, cadmium, chromium, lead, mercury, selenium, silver

Frequency of testing
"Several" since 1986

Ash handling and transportation

Ash handling
Combined (fly and bottom ash)

Ash covered in plant?
No

Ash covered while transported?
Yes

Mode of ash transportation
Leakproof, covered trucks

Ash treatment
None

Ash disposal
Oswego County Landfill at Volney

Monofill or codisposal
Codisposal

Landfill liners
2 liners

Leachate collection
Yes

Leachate treatment
Yes, at Fulton Water Treatment Plant

Expected life of landfill
Through 1988, expansion was anticipated

Length of ash disposal agreement
No agreement required; Oswego County owns ashfill

OTHER ENVIRONMENTAL FACTORS

Truck traffic

Number of trucks per day

Delivering municipal solid waste
60

Removing ash
5-7

Pass through residential areas?
Yes

Water management

Amount used per day
18,000 gallons per day

Used for
Quenching ash and cleaning plant

Source
Municipal well, boiler blow down

Disposal
Dirty water is sent to the Fulton Water Treatment Plant, cooling water is returned to the river untreated

ECONOMIC FACTORS

Costs

Capital construction (excluding financing)
$14.5 million (1986 dollars)

Operations per year (excluding ash management)
$1,600,000 (1988)

Revenues

Tipping fees (per ton)
None

Electricity (per kilowatt hour)
Not sold unless steam demand is low

Steam (per 1000 pounds)
$3.20

Scrap (per ton)
Not produced at plant

Control over revenues

Flow control?
Plant did not respond

Length of contract

Garbage
None

Electricity
None

Steam
15 years

PASCAGOULA

Pascagoula Energy Recovery Facility
5736 Elder Ferry Road
Moss Point, Mississippi 39501

Plant Characteristics

Owner/operator
City of Pascagoula/CFB Inc.

Vendor
Sigoure Frères of France

Start-up date
January 1985

Type
Mass burn

Type of fuel
Municipal solid waste, commercial and industrial waste, wood and #2 fuel oil for start-up

Energy products
Steam

Customer
Morton Thiokol

Energy rating
32,000 pounds per hour

Design capacity
150 tons per day

Capacity being used
120-125 tons per day

Worker Training and Experience

Number of employees
15

Training
Trained by manager and supervisor

Experience

Chief facility operator
On-the-job training

Shift supervisor
On-the-job training

Control room operator
On-the-job training

Service Area Characteristics

Area served
Pascagoula, Mississippi

Population of service area
70,000

Amount of municipal solid waste generated in service area
36,227 tons per year
75% residential (27,392 tons)
25% commercial (8,835 tons)

Solid waste management in area
50% landfilled
50% incinerated
No figures available as of 10/88 for % recycled

Materials collected for recycling
Some recovered metal at ash landfill

Planning

Criteria/method for sizing and design
Existing landfill records

Volume
Yes

Composition
Conducted in 1980-81

Lab analysis
Yes, to determine energy value of garbage

Siting

Nearest residence
2 miles

Location
Near steam customer; former wooded area

Citizen involvement in planning
Volunteers, citizens task force

Citizen opposition
Opposed the first site that was chosen, but agreed on current site

Plant Design and Operations

Garbage storage
Pit

Capacity (tons)
300

Capacity (days)
2

Screening of prohibited wastes

Technique
Visual

Responsible staff
Operators, furnace loaders

Materials prohibited
Oversized, bulky waste

Percent rejected
0.01%

Penalty type
Reprimand

Penalty enforced by
Plant personnel

Penalty ever levied?
Plant did not provide information

Furnace design

Loading technique
Continuous loading (overhead crane grapple)

Basic type
Modular dual-chambered combustor with excess air, rotary kiln

Modifications
None

Auxiliary burner
For start up and shut down

Operating temperature
1650°-1900°F

Automatic combustion controls
No

Emissions control equipment
Two-field electrostatic precipitator

Monitoring and Maintenance

Monitoring

Parameters monitored
Temperature, combustion efficiency, inlet and outlet temperature for air pollution control devices, steam pressure, opacity

Monitors connected to alarms?
No

Parameters recorded
Temperature, opacity

Time records kept
7 years

Frequency of reporting
As requested

Monitor calibration frequency
Every 2-3 months

Maintenance frequency

Furnace
21-day staggered shutdown schedule

Boiler
21-day staggered shutdown schedule

Stoker and grate
21-day staggered shutdown schedule

Air pollution control equipment
21-day staggered shutdown schedule

Turbine
21-day staggered shutdown schedule

Other sections
21-day staggered shutdown schedule

Air Emissions

Date of test(s)
December 1984

Test(s) conducted by
Environmental Monitoring Lab

Emissions tested for
Particulates, carbon monoxide, sulfur dioxide, oxides of nitrogen, hydrogen chloride, lead

Particulates
0.016, 0.019 grains per dry standard cubic foot at 12% CO_2 (two stacks)

Lead
0.12 pounds per hour (sum of two)

Carbon monoxide
111.5 pounds per hour (sum of two)

Hydrogen chloride
42.3 pounds per hour (sum of two)

Sulfur dioxide
17.1 pounds per hour (sum of two)

Oxides of nitrogen
18.0 pounds per hour (sum of two)

Ash

Ash amounts

Weight per day
25 tons

Volume per day
50 cubic yards

As % of original weight
25%

As % of original volume (estimated)
10%

Ash testing

Extraction Procedure Toxicity test

Materials tested for
Arsenic, barium, chromium, cadmium, lead, mercury, selenium, and silver

Frequency of testing
Every 3 months

Ash handling and transportation

Ash handling
Combined

Ash covered in plant?
No

Ash covered while transported?
No

Mode of ash transportation
Trucks

Ash treatment

Plant did not respond

Ash disposal

Monofill or codisposal
Monofill

Landfill liners
Clay liner

Leachate collection
Yes

Leachate treatment
To sanitary sewer treatment facility if it exceeds Extraction Procedure Toxicity test; if not, into surface water

Expected life of landfill
2005 (20 years from 1985)

Length of ash disposal agreement
2000 (15 years from 1985)

Other Environmental Factors

Truck traffic

Number of trucks per day

Delivering municipal solid waste
25-30

Removing ash
3

Pass through residential areas?
No

Water management

Amount used per day
Plant did not respond

Used for
Plant did not respond

Source
City of Moss Point

Disposal
Sump pump drains, water treated by Morton Thiokol

Economic Factors

Costs

Capital construction (excluding financing)
$6,800,000 (1985 dollars)

Operations per year (excluding ash management)
$960,945

Revenues

Tipping fees (per ton)
$16.83

Electricity (per kilowatt hour)
Not produced at plant

Steam (per 1000 pounds)
$2.00

Scrap (per ton)
Not produced at plant

Control over revenues

Flow control?

Plant did not respond

Length of contract

Garbage

15 years

Electricity

Not produced at plant

Steam

30 years

PIGEON POINT

The Delaware Electric Generating Facility (EGF)
The Delaware Reclamation Plant (DRP)
Pigeon Point, Delaware

PLANT CHARACTERISTICS

Owner/operator
EGF, General Electric Credit Corporation; DRP, Delaware Solid Waste Authority/ Raytheon Service

Vendor
EGF, Vicon Recovery; DRP, Raytheon Service

Start-up date
November 1987

Type
Refuse-derived fuel and/or mass burn

Type of fuel
Refuse-derived fuel (produced at DRP) and municipal solid waste

Energy products
Electricity, steam

Customer
Delmarca Power and Light, ICI Americas, Inc.

Energy rating
18 megawatts installed capacity, 13 megawatts operating capacity

Design capacity
EGF, 600 tons per day; DRP, 1000 tons per day municipal solid waste into 500 tons per day refuse-derived fuel, 350 tons per day sludge and organic fraction for composting

Capacity being used
EGF, 65% (1988); DRP, 89%

WORKER TRAINING AND EXPERIENCE

Number of employees
50

Training
Initiated as of 9/88

Experience

Chief facility operator
5 years minimum experience

Shift supervisor
5 years experience

Control room operator
5 years experience

SERVICE AREA CHARACTERISTICS

Area served
New Castle County

Population of service area
430,000 (1989 estimate)

Amount of municipal solid waste generated in service area
1800 tons per day
46% residential
54% commercial, industrial, and other

Solid waste management in area
Residential
30% landfilled (including ash)
50% incinerated
20% recycled

Materials recycled
Glass, aluminum, scrap metal, humus

PLANNING

Criteria/method for sizing and design

Volume
Estimates and weighing

Composition
Conducted in 1978 from samples at landfill next to the present plant

Lab analysis
Conducted in 1978

Siting

Nearest residence
0.5 miles

Location
Land adjacent to existing landfill was used; former dredge soil disposal area of the US Army Corps of Engineers

Citizen involvement in planning
Workshops/public hearings

Citizen opposition
No

PLANT DESIGN AND OPERATIONS

Garbage storage
Tipping floor

Capacity (tons)
EGF, 1500; DRP, 2000

Capacity (days)
EGF, 2.5; DRP, 2.5

Screening of prohibited wastes

Technique
Floor inspection; "jawboning" with collectors

Responsible staff
Floor inspector

Materials prohibited
Small drums, tires, propane, cylinders

Percent rejected
1-2%

Penalty type
Referred to Department of Natural Resources and Environmental Control

Penalty enforced by
Environmental Protection Officer

Penalty ever levied?
No

Furnace design

Loading technique
Batch loading using front end loader

Basic type
Modular furnace (dual chambered plus a tertiary chamber) with controlled air

Modifications
Plant did not respond

Auxiliary burner
Yes

Operating temperature
1800°F

Automatic combustion controls
Yes

Emissions control equipment
Three-field electrostatic precipitator, flue gas recirculation

MONITORING AND MAINTENANCE

Monitoring

Parameters monitored
Monitors not installed as of 1988; monitors for carbon monoxide, oxides of nitrogen, sulfur oxides, opacity, temperature, and steam pressure installed after 1988

Monitors connected to alarms?
All

Parameters recorded
Yes

Time records kept
Plant did not respond

Frequency of reporting
Plant did not respond

Monitor calibration frequency
Plant did not respond

Maintenance frequency

Furnace
Every 6 months

Boiler
Every 6 months

Stoker and grate
Every 6 months

Air pollution control equipment
Every 6 months

Turbine
Every 6 months

Other sections
Every 6 months

AIR EMISSIONS

Date of test(s)
December 1987

Test(s) conducted by
Roy F. Weston, Inc.

Emissions tested for

Particulates, carbon monoxide, sulfur dioxide, oxides of nitrogen, hydrogen chloride, 2,3,7,8-tetrachlorinated dibenzo-*p*-dioxin, arsenic, beryllium, cadmium, chromium, lead, mercury, nickel

Note When 4 numbers are given for a pollutant, they refer to separate tests of individual furnace/boiler units.

Particulates

0.003, 0.0015, 0.0064, 0.0053 grains per dry standard cubic foot at 12% O_2

Lead

2.94×10^{-5} grams per normal cubic meter at 7% O_2

Mercury

6.91×10^{-5} grams per normal cubic meter at 7% O_2

Carbon monoxide

5.5, 5.7, 6.7, 8.9 parts per million at 7% O_2

Dioxin/furan equivalents

0.508 nanograms per dry normal cubic meter (no correction factor)

Hydrogen chloride

587.09, 531.88, 488.21, 557.82 parts per million at 7% O_2

Sulfur dioxide

169.88, 162.39, 170.50, 152.80 parts per million at 7% O_2

Oxides of nitrogen

125.10, 104.47, 113.94, 116.54 parts per million at 7% O_2

Ash

Ash amounts

Weight per day

180 tons

Volume per day

Not measured

As % of original weight

20% without water

As % of original volume (estimated)

Not measured

Ash testing

Materials tested for

Dioxins, furans, metals

Frequency of testing

Eight times between 1986 and October, 1988

Ash handling and transportation

Ash handling

Combined (fly and bottom ash)

Ash covered in plant?

Yes

Ash covered while transported?

Yes

Mode of ash transportation

Covered trucks, not 100% leakproof

Ash treatment

No

Ash disposal

Cherry Island Landfill

Monofill or codisposal

Monofill

Landfill liners

One

Leachate collection

Yes

Leachate treatment

Sewage treatment

Expected life of landfill

2006 (17 years from 1989; expansion being planned)

Length of ash disposal agreement

21 years

Other Environmental Factors

Truck traffic

Number of trucks per day

Delivering municipal solid waste

250 to DRP

Removing ash

2

Pass through residential areas?

No

Water management

Amount used per day
Plant did not respond

Used for
Plant did not respond

Source
City water system

Disposal
Water used is recycled and the overflow is discharged to the sewers

ECONOMIC FACTORS

Costs

Capital construction (excluding financing)
EGF, $50 million; DRP, $72.3 million (1987 dollars)

Operations per year (excluding ash management)
EGF, $10.6 million; DRP, $10-11 million (includes debt service)
EGF, $4 million; DRP, $4-5 million (without debt service)

Revenues

Tipping fees (per ton)
$37.30

Electricity (per kilowatt hour)
$0.03- $0.05

Steam (per 1000 pounds)
Not sold

Scrap (per ton)
Not produced at plant

Control over revenues

Flow control?
No

Length of contract

Garbage
20 years

Electricity
Plant did not respond

Steam
None

TAMPA

The McKay Bay Refuse-to-Energy Facility
107 North 34th Street
Tampa, Florida 33605

Plant Characteristics

Owner/operator
City of Tampa/Wheelabrator Technologies

Vendor
Waste Management Energy Systems

Start-up date
September 1985

Type
Mass burn

Type of fuel
Municipal solid waste

Energy products
Electricity

Customer
Tampa Electricity Company

Energy rating
22.5 megawatts

Design capacity
1000 tons per day

Capacity being used
1000 tons per day; all capacity is being used, but 850 tons per day is the average for the past 3 years due to downtime for maintenance

Worker Training and Experience

Number of employees
20-24 (operations only)

Training
On-the-job, vendor training, stationary engineers license, 2-year in-house training course

Experience

Chief facility operator
Managed coal-fired plant

Shift supervisor
Two have incinerator experience, three have boiler experience

Control room operator
Plant did not respond

Service Area Characteristics

Area served
City of Tampa

Population of service area
285,200 in 1988

Amount of municipal solid waste generated in service area
365,000 tons per year
Plant did not respond for % residential or % commercial

Solid waste management in area
12% landfilled
88% incinerated
Pilot recycling program initiated as of 9/88

Materials collected for recycling
Newspapers, bottles, cans, yard waste

Planning

Criteria/method for sizing and design

Volume
Volume estimate based on population and employment, and on historical tonnage records

Composition
Waste composition, study

Lab analysis
None; estimates based on national studies

Siting

Nearest residence
1/2 mile

Location
Site of previous incinerator

Citizen involvement in planning
Citizens advisory committee

Citizen opposition
None

Operating temperature
1850°-2000°F

Automatic combustion controls
Yes (Philips control system)

Emissions control equipment
Two-field electrostatic precipitator

PLANT DESIGN AND OPERATIONS

Garbage storage
Pit

Capacity (tons)
2000

Capacity (days)
2

Screening of prohibited wastes

Technique
Crane operator, random truck search, waste screening questionnaire

Responsible staff
Crane operator, city representative, plant operations manager

Materials prohibited
Explosives, hazardous waste, radioactive waste, demolition debris, tree stumps, large machinery, sewage sludge, liquid waste

Percent rejected
Less than 1% (approximately 300 tons per year)

Penalty type
Prohibited waste returned to generator

Penalty enforced by
City of Tampa, tipping floor attendant

Penalty ever levied?
Yes; recent revision to city code allows city to bill responsible party for cleanup of prohibited waste

Furnace design

Loading technique
Continuous loading

Basic type
Rotary kiln with excess air

Modifications
Plant did not respond

Auxiliary burner
None

MONITORING AND MAINTENANCE

Monitoring

Parameters monitored
Temperature, oxygen, opacity, steam pressure, inlet/outlet temperatures of air pollution control devices

Monitors connected to alarms?
Yes

Parameters recorded
All parameters recorded on plant data logging system

Time records kept
2 years

Frequency of reporting
4 times a year

Monitor calibration frequency
Quarterly

Maintenance frequency
Each of the four units is taken out of service once a year for scheduled maintenance. Each unit is out of service for about one week every fourth month.

Furnace
A few days a month

Boiler
A few days a month

Stoker and grate
A few days a month

Air pollution control equipment
A few days a month

Turbine
Five years plus overhaul schedule

Other sections
As needed

AIR EMISSIONS

Date of test(s)
October 1987, unless otherwise noted

Test(s) conducted by
Environmental Engineering Consultants, Inc.

Emissions tested for
Particulates, carbon monoxide, sulfur dioxide, oxides of nitrogen, fluorides, volatile organic compounds, beryllium, lead, mercury

Particulates
0.012 grains per dry standard cubic foot at 12% CO_2

Lead
0.4 pounds per hour = 7.76×10^{-4} grams per normal cubic meter at 12% CO_2 (tested 9/85)

Mercury
0.36 pounds per hour = 9.31×10^{-4} grams normal cubic meter at 12% CO_2 (tested 9/85)

Carbon monoxide
21.9 pounds per hour

Sulfur dioxide
79.7 pounds per hour = 78.8 parts per million at 12% CO_2

Oxides of nitrogen
94.8 pounds per hour

ASH

Ash amounts

Weight per day
170 tons per day

Volume per day
Unknown

As % of original weight
18-19% (dry weight)

As % of original volume (estimated)
5%

Ash testing
Extraction Procedure Toxicity test

Materials tested for
Arsenic, barium, cadmium, chromium, lead, mercury, selenium, silver

Frequency of testing
Many tests have been done in conjunction with ash "reuse" investigations, annual Extraction Procedure Toxicity test required by landfill and facility permits

Ash handling and transportation

Ash handling
Combined (fly and bottom ash)

Ash covered in plant?
No

Ash covered while transported?
Yes

Mode of ash transportation
Leakproof, covered trucks

Ash treatment
None

Ash disposal

Monofill or codisposal
Codisposal

Landfill liners
One

Leachate collection
Yes

Leachate treatment
Leachate taken to county wastewater treatment facility

Expected life of landfill
2008 (20 years from 1988)

Length of ash disposal agreement
No contract

OTHER ENVIRONMENTAL FACTORS

Truck traffic

Number of trucks per day

Delivering municipal solid waste
150

Removing ash
10-15

Pass through residential areas?
No

Water management

Amount used per day
460,000 gallons per day

Used for
Cooling tower

Source
Advanced wastewater treatment plant

Disposal
Same as source; small amount of potable water is used for boiler makeup

ECONOMIC FACTORS

Costs

Capital construction (excluding financing)
$70,000,000 (1985 dollars)

Operations per year (excluding ash management)
$4,500,000

Revenues

Tipping fees (per ton)
$18.00 (operator), $58.00 (commercial)

Electricity (per kilowatt hour)
$0.026

Steam (per 1000 pounds)
Not sold

Scrap (per ton)
$0-$10

Control over revenues

Flow control?
Yes

Length of contract

Garbage
20 years

Electricity
21 years

Steam
None

TULSA

Walter B. Hall Resource Recovery Facility
2122 South Yukon Avenue
Tulsa, Oklahoma 74107

Plant Characteristics

Owner/operator
Ogden Martin Systems of Tulsa, Inc.

Vendor
Ogden Martin Systems of Tulsa, Inc.

Start-up date
March 1986

Type
Mass burn

Type of fuel
Municipal solid waste (residential and commercial) and light industrial waste

Energy products
Steam and electricity

Customer
Steam to Sun Refining & Marketing Co., electricity to Public Service Co. of Oklahoma

Energy rating
18.2 megawatts

Design capacity
1125 tons per day at 4500 British thermal units per pound

Capacity being used
925 tons per day

Worker Training and Experience

Number of employees
40

Training
Classroom and on-the-job training

Experience

Chief facility operator
Tennessee Valley Authority operator training and 7 years of waste-to-energy experience

Shift supervisor
Three years as control room operator

Control room operator
Three years experience

Service Area Characteristics

Area served
City of Tulsa (metropolitan Tulsa)

Population of service area
380,000 (1990 estimate)

Amount of municipal solid waste generated in service area
% residential and commercial not measured

Solid waste management in area
35-40% landfilled
60-65% incinerated
As of 7/90, no recycling program

Materials collected for recycling
Ferrous metal from ash stream

Planning

Criteria/method for sizing and design

Volume
Projections for waste generation in Tulsa County; scale records obtained from two operating landfills

Composition
None

Lab analysis
None

Siting

Nearest residence
5 blocks

Location
Near steam customer; former vacant land zoned for industrial use, formerly used for cattle grazing

Citizen involvement in planning
Open meetings

Citizen opposition
None

Plant Design and Operations

Garbage storage
Pit

Capacity (tons)
3350

Capacity (days)
3

Screening of prohibited wastes

Technique
Random sampling by dumping incoming loads on floor

Responsible staff
City personnel/Ogden Martin personnel

Materials prohibited
Noncombustible construction material, demolition debris, bulky noncombustibles, hazardous wastes, sewage sludge, white goods, motor vehicles

Percent rejected
0.5%

Penalty type
Reloading on truck, paying for load plus landfill cost

Penalty enforced by
Solid Waste Management Department of City of Tulsa, Tulsa Authority for Recovery of Energy

Penalty ever levied?
Three times (as of July 1990)

Furnace design

Loading technique
Continuous loading by cranes

Basic type
Single-chambered waterwall furnace with excess air

Modifications
Third identical combustion unit added September, 1987

Auxiliary burner
No

Operating temperature
1745°F (minimum design)

Automatic combustion controls
Yes

Emissions control equipment
Three-field electrostatic precipitator

Monitoring and Maintenance

Monitoring

Parameters monitored
Temperature, oxygen, carbon monoxide, steam pressure

Monitors connected to alarms?
Yes

Parameters recorded
Same as monitored

Time records kept
5 years

Frequency of reporting
Not required

Monitor calibration frequency
Annually

Maintenance frequency

Furnace
Approximately 6-month intervals for cleaning

Boiler
Approximately 6-month intervals for cleaning

Stoker and grate
Approximately 6-month intervals for cleaning

Air pollution control equipment
Approximately 6-month intervals for cleaning

Turbine
Plant did not respond

Other sections
As required or 1/2 year intervals

Air Emissions

Test(s) conducted by
Ogden Projects, Inc.

Date of test(s)
July, October 1986

Emissions tested for

Particulates, total polychlorinated dibenzodioxins/polychlorinated dibenzofurans, 2,3,7 8-tetrachlorinated dibenzo-*p*-dioxin, carbon monoxide, hydrogen chloride, sulfur dioxide, sulfuric acid, oxides of nitrogen, fluorides, hydrogen fluoride, volatile organic compounds, beryllium, lead, mercury

Note When 2 numbers are given for a pollutant, they refer to separate tests of each of the 2 furnace/boiler units.

Particulates

0.0049, 0.0095 grains per dry standard cubic foot at 12% CO_2

Lead

0.000415 grams per normal cubic meter at 12% CO_2

Mercury

0.000419 grams per normal cubic meter at 12% CO_2

Carbon monoxide

17, 21 parts per million dry volume (no correction)

Dioxin/furan equivalents

1.735 nanograms per normal cubic meter at 12% CO_2

Hydrogen chloride

422, 402 parts per million at 12% CO_2

Sulfur oxides

94.9, 97.1 parts per million at 12% CO_2

Oxides of nitrogen oxides

361, 372 parts per million dry volume at 7% O_2

Ash

Ash amounts

Weight per day

235 tons per day

Volume per day

Not available

As % of original weight

25-26%

As % of original volume (estimated)

5-7%

Ash testing

Extraction Procedure Toxicity test

Materials tested for

Arsenic, barium, cadmium, chromium, lead, mercury, selenium, silver

Frequency of testing

Further testing not required

Ash handling and transportation

Ash handling

Combined (fly and bottom ash)

Ash covered in plant?

Yes; covered conveyors and ash handling building

Ash covered while transported?

Yes

Mode of ash transportation

Leakproof, covered trucks

Ash treatment

None

Ash disposal

North Tulsa Landfill

Monofill or codisposal

Codisposal

Landfill liners

None

Leachate collection

None

Leachate treatment

None

Expected life of landfill

1990 (less than a year from December 1989)

Length of ash disposal agreement

Plant did not respond

Other Environmental Factors

Truck traffic

Number of trucks per day

Delivering municipal solid waste

144-166

Removing ash

17

Pass through residential areas?

No

Water management

Amount used per day

Varies, depending on amount of steam produced

Used for

Quenching ash, boiler blowdown, steam makeup, sanitizing

Source

City water

Disposal

Sanitary sewer

Economic Factors

Costs

Capital construction (excluding financing)

$76 million (1986 dollars)

Operations per year (excluding ash management)

$4,500,000 (1986) plus escalation and pass through costs

Revenues

Tipping fees (per ton)

$21.00

Electricity (per kilowatt hour)

$0.02

Steam (per 1000 pounds)

$2.75

Scrap (per ton)

Not produced at plant

Control over revenues

Flow control?

Yes

Length of contract

Garbage

20 years

Electricity

20 years

Steam

20 years

WESTCHESTER

Westchester County Refuse Energy System Company (RESCO)
One Charles Point Avenue
Peekskill, New York 10566

Plant Characteristics

Owner/operator
RESCO Company L.P./Westchester RESCO Operating Company

Vendor
Wheelabrator Technologies

Start-up date
October 1984

Type
Mass burn

Type of fuel
Municipal solid waste (residential and commercial)

Energy products
Electricity

Customer
Con Edison of New York

Energy rating
60 megawatts

Design capacity
2250 tons per day

Capacity being used
1800 tons per day

Worker Training and Experience

Number of employees
70

Training
On-the-job training

Experience

Chief facility operator
Boiler operating experience

Shift supervisor
Boiler operating experience

Control room operator
Boiler operating experience

Service Area Characteristics

Area served
Westchester County

Population of service area
850,000 (as of 1988)

Amount of municipal solid waste generated in service area
2472 tons per day
60% residential
40% commercial

Solid waste management in area
As of October 1988
21% landfilled
73% incinerated
6% recycled

Materials collected for recycling
Returnable beverage containers, paper

Planning

Criteria/method for sizing and design

Volume
County waste data for residential; comprehensive assessment for commercial waste

Composition
Method not supplied

Lab analysis
None

Siting

Nearest residence
1/2 - 3/4 mile

Location
New

Citizen involvement in planning
Public meeting

Citizen opposition
From local residents in Sprain Ridge and Yonkers, none in Peekskill

Plant Design and Operations

Garbage storage
Pit

Capacity (tons)
6000

Capacity (days)
3

Screening of prohibited wastes

Technique
Visual

Responsible staff
Plant did not respond

Materials prohibited
Hazardous waste, hospital waste, industrial wastes, demolition debris, oversized items

Percent rejected
Less than 1%

Penalty type
None

Penalty enforced by
Not applicable

Penalty ever levied?
Not applicable

Furnace design

Loading technique
Continuous loading

Basic type
Single-chambered waterwall furnace with excess air

Modifications
Plant did not respond

Auxiliary burner
Yes

Operating temperature
More than 2000°F

Automatic combustion controls
Steam, oxygen in the flue

Emissions control equipment
Three-field electrostatic precipitator

Monitoring and Maintenance

Monitoring

Parameters monitored
Temperature, oxygen, carbon monoxide, opacity, oxides of nitrogen, steam pressure

Monitors connected to alarms?
Opacity, sulfur dioxide, carbon monoxide

Parameters recorded
Same as monitored

Time records kept
3-5 years

Frequency of reporting
Not reported

Monitor calibration frequency
Daily

Maintenance frequency

Furnace
Every 3-4 months

Boiler
Every 3-4 months

Stoker and grate
Every 3-4 months

Air pollution control equipment
Every 3-4 months

Turbine
Once every 5 years

Other sections
Plant did not respond

Air Emissions

Date of test(s)
1988

Test(s) conducted by
New York State Department of Environmental Conservation

Emissions tested for
Particulates, 2,3,7,8-tetrachlorinated dibenzo-*p*-dioxin, hydrogen chloride, sulfur dioxide, oxides of nitrogen, arsenic, beryllium, cadmium, chromium, lead, manganese, mercury, nickel, vanadium, zinc

Particulates
0.016 grains per dry standard cubic foot at 7% O_2

Lead
1.5×10^{-4} grams per dry normal cubic meter at 7% O_2

Mercury
1.92×10^{-3} grams per dry normal cubic meter at 7% O_2

Dioxin/furan equivalents
4.16 nanograms per dry normal cubic meter at 7% O_2

Hydrogen chloride
646 parts per million at 7% O_2

Sulfur dioxide
140 parts per million at 7% O_2

Oxides of nitrogen
240 parts per million at 7% O_2

Ash

Ash amounts

Weight per day
461 tons per day

Volume per day
Plant did not respond

As % of original weight
25%

As % of original volume (estimated)
5%

Ash testing
Extraction Procedure Toxicity test, Toxic Characteristics Leaching Procedure

Materials tested for
Arsenic, barium, cadmium, chromium, lead, mercury, selenium, silver

Frequency of testing
Twice a year

Ash handling and transportation

Ash handling
Combined (fly and bottom ash)

Ash covered in plant?
No

Ash covered while transported?
Yes

Mode of ash transportation
Covered truck

Ash treatment
Fly ash treated by agglomeration and spray wetting

Ash disposal
Sprout Brook Landfill

Monofill or codisposal
Monofill

Landfill liners
One

Leachate collection
Yes

Leachate treatment
Sewage treatment plant

Expected life of landfill
2010 (22 years from 1988)

Length of ash disposal agreement
2004 (20 years from 1984)

Other Environmental Factors

Truck traffic

Number of trucks per day

Delivering municipal solid waste
150-250

Removing ash
25

Pass through residential areas?
Not on route to plant

Water management

Amount used per day
Plant did not respond

Used for
Quenching ash

Source
Municipal water supply

Disposal
If it meets standards, into city sewer; if not, reused in plant or hauled to a publicly owned treatment facility

Economic Factors

Costs

Capital construction (excluding financing)
$239 million (1984 dollars)

Operations per year (excluding ash management)
Plant did not respond

Revenues

Tipping fees (per ton)
$17.00

Electricity (per kilowatt hour)
Plant did not respond

Steam (per 1000 pounds)
Not sold

Scrap (per ton)
Not separated from ash

Control over revenues

Flow control?
Yes

Length of contract

Garbage
25 years

Electricity
25 years

Steam
None

APPENDIX B: METHODOLOGY

INFORM studied incineration of municipal solid waste in the United States in order to answer two basic questions:

1. What is state-of-the-art incineration? What technologies and planning and operating practices lead to the cleanest possible incineration?
2. To what extent do 15 waste-to-energy plants — selected to illustrate the diversity of technologies and other factors — achieve this state of the art?

To answer these questions, we identified state-of-the-art technologies and practices; established state-of-the-art emissions levels for six key air pollutants; examined design features, operating practices, and environmental performance at a diverse cross section of modern plants; and compared the 15 plants to each other and to the state of the art.

INFORM's information about state-of-the-art practices and technologies came from a careful review of the available literature on existing plants. Our information about the individual incinerators came from field visits to the plants, interviews with facility managers, emissions test reports, and follow-up questionnaires.

Defining the State of the Art

To define state-of-the-art technologies and practices, and to identify state-of-the-art emissions levels, INFORM examined the literature on the performance of waste-to-energy incinerators in the United States, Canada, and Europe. We looked at permits for existing plants; recent emissions test reports from operating incinerators worldwide; regulations and recommendations of environmental authorities, including the United States Environmental Protection Agency, the Swedish Environmental Protection Board, and Environment Canada; and technical papers from professional conferences, proceedings, and journals.

Most of the information on the state of the art is covered in Chapter 3, "The Technology of Garbage Burning." Where appropriate, the source is footnoted. In addition, all sources are listed in Appendix C, "Bibliography." Two documents deserve special mention because they each contain extensive bibliographies. *Improving the Environmental Performance of MSW Incinerators* and *Technologies for Minimizing the Emission of NO_x from MSW Incinerators,* both prepared for INFORM by Marjorie J. Clarke, refer to emissions tests from more than 50 incinerators, from 1970 through 1988.

The one essential criterion for the state-of-the-art emissions levels INFORM identified was that they have been achieved, with regularity, in actual practice. Thus, these levels are in fact conservative, since in many cases substantially lower emissions levels are already being achieved in test situations. The specific sources for establishing the state-of-the-art level for emissions of each of the six key pollutants (particulates, carbon monoxide, hydrogen chloride, sulfur oxides, dioxins/furans, and oxides of nitrogen) are listed in Table 3-4, "State-of-the-Art Emissions Levels."

It should be stressed that the state of the art is always changing and improving. The technologies, practices, and emissions levels discussed in this book are likely to continue to improve. Further, it should be stressed that the current state of the art goes well beyond what United States laws, regulations, and permits now require.

Selection of the Study Plants

INFORM chose to examine 15 waste-to-energy incinerators representing a cross section of manufacturers, sizes, geographical locations, furnace designs, emissions control equipment, operating practices, and regulatory and economic environments. Since we wanted to assess relatively up-to-date facilities, we picked plants that had started to operate between 1981 and 1987. We polled industry experts, including managers of garbage-burning plants and government regulators, to ensure that our selection of study plants was broadly representative of an industry with rapidly changing technologies and practices.

As a first step, INFORM selected three plants for a pilot study. These plants, Albany, Pascagoula, and Westchester, were chosen because they varied in their size, ownership, fuel, and energy product. The information obtained in the pilot phase helped INFORM sharpen the focus of the full study.

The individual plants were selected for the reasons listed in Table B-1.

Field Research and Follow-Up

Having defined state-of-the-art technologies and practices and identified state-of-the-art emissions levels for six key air pollutants, INFORM turned to the in-depth evaluation of the 15 selected study plants. INFORM researchers visited each of the waste-to-energy incinerators, interviewed plant managers and others familiar with the plants, sent follow-up questionnaires to obtain additional information, examined air emission test reports, and verified the accuracy of the data obtained with the plant managers. The analysis of these data is INFORM's alone.

Plant Visits

Each plant visit lasted one or two days and consisted of interviews with plant managers and municipal officials, as well as a tour of the plant itself. Each interview with a plant manager followed a standard outline and included a standard series of topics.

Table B-1 Selection of Study Plants

State Plant	Reason for Selection
California	
Commerce	Only operational plant in the United States with ammonia injection process to remove oxides of nitrogen from the flue gases.
Delaware	
Pigeon Point	Dual-chambered furnace; flue gas recirculation; vendor of furnace equipment (Vicon Recovery) broadened array of manufacturers included in study.
Florida	
Dade County	Very large (3000 ton per day) design capacity; plant operating while undergoing reconstruction.
Lakeland	Plant is basically a coal-burning power plant, with 10 percent refuse-derived fuel.
Tampa	Rotary kiln furnace broadened array of technologies included in study.
Maine	
Auburn	An early (1981) Consumat plant; dual-chambered furnace; operated with no functioning add-on emissions control equipment; plant closed in February 1990 (after research for this study was completed).
Biddeford/Saco	Privately owned and operated refuse-derived fuel plant with scrubber and fabric filter; broadened array of technologies and ownership arrangements included in study.
Maryland	
Baltimore	Large design capacity; publicly owned and privately operated; plant design similar to that of Westchester; chosen to assess differences in performances of similar plants.
Mississippi	
Pascagoula	Small, publicly owned and privately operated mass burn plant selling steam to private industry; dual-chambered, rotary kiln furnace on a vertical axis; one of three plants in pilot study.
New Hampshire	
Claremont	Small privately owned and operated mass burn plant with scrubber and fabric filter.

(continued on next page)

Table B-1 Selection of Study Plants (cont'd.)

State Plant	Reason for Selection
New York	
Albany	Medium sized, state-owned, refuse-derived fuel plant selling steam to the state; one of three plants in pilot study.
Oswego	Dual-chambered mass burn plant, built in 1986, with Consumat technology; enabled comparison with older Auburn plant.
Westchester	Large, mass burn plant, public/private joint venture, selling electricity to local utility; one of three plants in pilot study.
Oklahoma	
Tulsa	Medium-sized, mass burn plant; built and operated by a major incinerator vendor (Ogden Martin).
Oregon	
Marion County	First United States plant with combination of fabric filter and acid gas scrubber.

Interview Outline for Plant Visits

I. Planning for construction
  A. Reason for building the plant
  B. Choice of site
  C. Determination of plant size
  D. Vendor/manufacturer/operator
  E. Analysis of waste
  F. Permitting process
II. Construction
  A. Ownership and management
  B. Cost of construction
  C. Start-up date
III. The plant
  A. Waste sorting and screening
  B. Tip floor operation
  C. Monitors and control room
  D. Boiler operation
  E. Emissions control devices
  F. Ash management and disposal
  G. Water use
  H. Truck traffic
  I. Maintenance schedules
  J. Markets for electricity, steam, and other products
  K. Costs of operation
IV. Worker safety and training
V. Testing
  A. Emissions tests
  B. Ash tests
VI. Community
  A. Opposition to plant
  B. Involvement in planning process
VII. Regulatory environment
  A. Local
  B. State
  C. National

Follow-Up Research

INFORM confirmed the information collected during the on-site visits and obtained additional data through extensive telephone interviews with plant staff, state and local officials, and other individuals knowledgeable about the operations of each incinerator. In addition, INFORM sent a follow-up questionnaire to the manager of each plant; the format of the plant profiles in Appendix A is based on the format of these questionnaires. The manager of every plant except the Albany incinerator responded, either verbally or in writing.

Air Emissions Data

INFORM obtained air emissions test data from a variety of sources. In some cases, the plants provided test reports; in others, the results of tests were published by state authorities. The tests themselves were largely carried out by private companies under contract to either the plant or the state; the plant profiles indicate the testing organizations. When necessary, as discussed more fully in the section on "Standardizing Procedures," INFORM translated data from the test reports into a standard form in order to permit comparison of results from different plants.

It is important to note that stack emissions data have certain inherent limitations. In particular, while they provide a snapshot of an incinerator's performance at a specific time, the data may or may not be representative of the plant's "typical" emissions levels. Further, they do not provide a picture of an incinerator's operation over time.

Plant Profiles

As a final step, in late 1990, INFORM compiled all the information obtained through plant visits, interviews, questionnaires, and review of emissions test data into individual plant profiles for each facility and sent them to the plant managers for verification of the facts before publication. The cover letter indicated that INFORM would assume that the information in the profiles was accurate unless the incinerator managers indicated any corrections.

Eleven of the fifteen plant managers reviewed and returned the profiles. Despite repeated requests from INFORM, the managers of the Albany plant and the three plants operated by Wheelabrator (Baltimore, Claremont, and Westchester) did not respond. Thus, every plant manager had ample opportunity to verify the data used throughout this book, in both the text analysis and the plant profiles in Appendix A.

Comparing Air Emissions Data

In order to be able to compare the environmental performance of individual waste-to-energy incinerators to each other and to the state of the art, emissions levels must be expressed in a uniform format, using the same measurement units and corrected to a standard set of operating conditions. For the purposes of this study, INFORM used 7 percent oxygen, or its rough equivalent of 12 percent carbon dioxide, as its standard operating conditions. Measurement units depended on the individual pollutant: grains per dry standard cubic foot for particulates; grams per dry normal cubic meter for lead and mercury; grams per dry normal cubic meter, Eadon toxic equivalents, for dioxins/furans; and parts per million for carbon monoxide, hydrogen chloride, sulfur dioxide, and oxides of nitrogen.

When the data in the emissions test reports did not conform to these standard conditions or measurement units, INFORM tried to convert the figures using standard calculations. In some cases, however, it was not possible to standardize the data. For example, INFORM could not convert dioxins and furans measured as total emissions to Eadon toxic equivalents because the conversion is complex and the basic data necessary for the conversion were not provided. In these cases, the plant profile lists the figures provided by the plant but the data are not included in the text analysis; text tables indicate whether data are not included because the plant did not provide them or because the figures could not be converted to a comparable format.

In some cases, plants provided reports of emissions tests carried out at different times. INFORM used the figures that were most recent at the time of the original data collection. Although some plants provided newer emissions data during the final check of the plant profiles, INFORM did not use these figures because not all the plants did so. The plant profiles show the dates of the tests used for this study.

Stack testing methodology usually involves several observations during one day or a series of observations over several days. In some cases, the variances between observations are minor; in others great. Test reports may present these multiple observations as discrete numbers or as averages. Where discrete numbers were provided, INFORM averaged them to facilitate comparison.

Additionally, many plants consist of more than one combustion unit (furnace and boiler), each of which, frequently, has its own emissions stack. During the stack test process, emissions from each stack are measured separately (although not all stacks may be tested during any one test process). Test reports may present the individual stack reports separately, or may combine them through addition or averaging. When separate reports were provided, INFORM averaged them for the analysis in the text and tables in Chapter 4, but lists individual stack results in the plant profiles in Appendix A.

Standardizing Procedures

When calculations were required to standardize emissions data, INFORM used the following formulas (temperature, pressure and humidity conversions were not attempted):

To convert from pounds per hour to grams per normal cubic meter:

Z grams per normal cubic meter = [(Y pounds per hour) ÷ (Flow in dry standard cubic feet per minute × 60)] × 453.6 grams per pound ÷ 0.0283 cubic meters per cubic foot.

$$Z\ g/nm^3 = \frac{Y\ lb/hr}{F\ (dsf^3/minute) \times (60\ min.\ /hour)} \times \frac{453.6\ grams/pound}{0.0283\ m^3/f^3}$$

To convert from grams per normal cubic meter at a given correction to parts per million at the same correction:

The value in parts per million = the value in milligrams per normal cubic meter × 24.5 ÷ the molecular weight of the substance.

$$Z\ ppm = Y\ milligrams\ /nm^3 \times \frac{24.5}{MW}$$

Converting from standard cubic feet to parts per million requires an additional conversion from cubic feet to cubic meters (1 cubic foot equals 0.0283 cubic meters). The standard temperature is either 20°C or 70°F; these are sufficiently equivalent so as not to matter in the conversion.

To convert from grams per normal cubic meter to grains per dry standard cubic foot:

$$\text{g/nm}^3 = 0.437 \text{ gr/dsf}^3$$

There are 7000 grains per pound. There are 0.02832 cubic meters per cubic foot. There are 453.6 grams per pound.

$$(7000 \times 0.02832)/453.6 = 0.437$$

To standardize the measurement for oxygen:

The value in grams per normal cubic meter (or any concentration of the pollutant in question) at 7 percent oxygen = the value in grams per normal cubic meter at the percent of oxygen measured concurrent with the stack test times 14 ÷ (21 - the measured percent of oxygen).

$$Z \text{ g/nm}^3 \text{ at } 7\%O_2 = Y \text{ g/nm}^3 \times \frac{14}{21 - \text{measured } \%\ O_2}$$

To standardize the measurement for carbon monoxide:

The value in grams per normal cubic meter (or any concentration parameter of the pollutant in question) at 12 percent carbon dioxide = (12 ÷ the measured percent of carbon monoxide) × the value in grams per normal cubic meter at the measured percent of carbon monoxide.

$$Z \text{ g/nm}^3 \text{ at } 12\%\ CO_2 = Y \text{ g/nm}^3 \times \frac{12}{\text{measured } \%\ CO_2}$$

Notes on the 15 Plants

The air emissions data obtained for each plant are indicated in each plant profile, along with the date(s) of the tests. The comments that follow address any noteworthy aspects of either the source or the standardization of the data.

California

Commerce (test dates: May/June 1987). The data contained in the test reports from Commerce were corrected to 3% O_2. INFORM converted the data to 7% O_2.

Delaware

Pigeon Point (test date: December 1987). INFORM converted the data provided for gaseous emissions (carbon monoxide, hydrogen chloride, and sulfur dioxide) to 7% O_2 and averaged the results of three tests. The plant profile lists individual averages for each of the four units of the plant for these emissions and for particulates; these four units were averaged for the analysis in Chapter 4.

For dioxins and furans, INFORM took data about total equivalent 2,3,7,8-tetrachlorodibenzodioxin (2,3,7,8-TCDD) from Table 11 in an October 18, 1988 letter from the testing company to N. C. Vasuki, the General Manager of the Delaware Solid Waste Authority, and considered these data to correspond to Eadon toxic equivalents. However, since there was no correction for either O_2 or CO_2, INFORM could not compare the dioxin/furan emissions to those of other plants.

Florida

Dade County (test date: April 1987). Although Dade County has four units, one unit was not operating at the time of particulate testing due to the reconstruction process. The data presented to INFORM reflects an average of emissions from the three operating boilers (boilers 2, 3, and 4).

An additional particulate test on boiler 2 was carried out in January 1988 after the addition of a third field to the electrostatic precipitator; the emissions for that one boiler improved to 0.0043 grains per dry standard cubic foot from the earlier 0.0258 grains per dry standard cubic foot average for the three boilers. INFORM did not use this test result in the text because our analysis considers the Dade County plant before the retrofitting process was completed.

Lakeland (test date: April 1988). Test reports gave particulates and sulfur dioxide in pounds per million BTUs. INFORM converted the data to, respectively, grains per dry standard cubic foot and parts per million, both at 7% O_2.

Tampa (test date: October 1987). Test reports gave emissions of all pollutants tested except particulates in units of pounds per hour. INFORM was able to convert the data for lead and mercury to grams per normal cubic meter and for sulfur dioxide to parts per million (all at 12% CO_2). However, the plant did not provide sufficient information to allow INFORM to convert the data for carbon monoxide and oxides of nitrogen to a format that would permit comparison with data from other plants.

Since research for this study was completed, an October 1989 emissions test became available. It shows particulate emissions of 0.009 grains per dry standard cubic foot (12% O_2), a level slightly improved from the 0.012 grains per dry standard cubic foot figure used in this study. It is interesting to note that the same report shows substantial increases in the emissions of oxides of nitrogen (from 135.8 pounds per hour in October 1987 to 230.7 pounds per hour) and of sulfur dioxide (from 79.7 pounds per hour to 111.6 pounds per hour); as explained above, INFORM did not have enough information to convert these figures to a standardized format. To maintain comparability among the plants in this study, INFORM used only the 1987 data in the analysis.

Maine

Auburn (test date: October 1982). The October 1982 test is the most recent one.

Biddeford/Saco (test dates: September/November/December 1987). INFORM averaged data from individual tests. The data on dioxin and furan toxic equivalents come from "Results of Emissions and Ash Testing at the Maine Energy Recovery Company Waste-to-Energy Plant," by Francis A. Ferraro and Randall J. Parenteau, dated June

1988, prepared for presentation at the 81st Annual Meeting of the Air Pollution Control Association in Dallas, Texas. The authors of that report calculated toxic equivalents from data provided by the plant.

Maryland

Baltimore (test date: January 1985). The test report gives particulate emissions for three furnace/boiler units; INFORM averaged these figures. INFORM standardized the data for carbon monoxide, oxides of nitrogen, and sulfur dioxide to 7% O_2 and averaged individual figures provided for three furnace/boiler units.

Mississippi

Pascagoula (test date: December 1984). For particulate emissions, INFORM averaged the results of emissions tests on each of the plant's two combustion units. The test report gives emissions for the other pollutants measured (carbon monoxide, hydrogen chloride, sulfur dioxide, and oxides of nitrogen) in units of pounds per hour; INFORM did not have sufficient data to be able to convert these figures to a standardized format.

A 1990 test, not used for this study, showed a substantial increase in particulate emissions (from 0.018 grains per dry standard cubic foot to 0.031 grains per dry standard cubic foot, both at 12% CO_2) and a substantial decrease in carbon monoxide emissions (from 111.5 pounds per hour to 2.02 pounds per hour, not convertible to standard format).

New Hampshire

Claremont (test dates: May/October 1987). INFORM corrected the test report data for carbon monoxide, hydrogen chloride, sulfur dioxide, and oxides of nitrogen to 7% O_2, and averaged the results of three test runs for each of two combustion units. The plant profile lists the average for each unit for each pollutant, and the text analysis uses averages of the results from both units. The figure for particulates also represents an average of emissions from each of the two units. The plant tested for total dioxins and furans but did not provide toxic equivalent figures.

New York

Albany (test dates: July 1984/March 1987). Data taken from New York State Draft Environmental Impact Statement (DEIS) for waste-to-energy incinerators, attached to proposed revisions of the state's Part 219 regulations, 1988. The DEIS reported particulate emissions of 0.139 grains per dry standard cubic foot, based on a 1984 test; that did not pass the plant's permit conditions of 0.08 grains per dry standard cubic foot. The plant then modified its furnace system with improved combustion controls and was retested by the state Department of Environmental Control in December 1986, March 1987, and September 1988. INFORM used the data from the most recent complete test, in March 1987, for particulate emissions (0.020 grains per dry standard cubic foot).

Oswego (test dates: June/August/September 1986). INFORM used an average of the September 1986 tests; only one of four furnace/boiler units was tested.

Westchester (test dates: September/October 1986). Data taken from New York State

Draft Environmental Impact Statement (DEIS) for waste-to-energy incinerators, attached to proposed revisions of the state's Part 219 regulations, 1988.

Oklahoma

Tulsa (test dates: July/October 1986). INFORM averaged emissions data from the plant's two combustion units; the data were provided by the plant management (Ogden Martin). INFORM could not use the data on carbon monoxide emissions because not enough information was provided to permit standardization of the figures to either 7% O_2 or 12% CO_2.

Oregon

Marion County (test dates: September/October 1986). INFORM averaged emissions data from the plant's two combustion units; the data were provided by the plant management (Ogden Martin).

Appendix C. Bibliography

Ahmed, A. K., and F. Perera. *Respirable Particles.* 1979.

Air and Waste Management Association/US Environmental Protection Agency. Conference Papers and Abstracts from the Second International Specialty Conference on Municipal Waste Combustion, Tampa, FL: April 15-19, 1991.

California Air Resources Board. "Air Pollution Control at Resource Recovery Facilities." May, 1984; revised, 1991.

Environment Canada. "Review of Japanese Incinerator Technology," by J. Pohl. International Workshop on Municipal Waste Incineration, October 1-2, 1987. National Incinerator and Evaluation Program.

Environment Canada. "The National Incinerator Testing and Evaluation Program: Environmental Characterization of Mass Burning Incinerator Technology at Quebec City." June, 1988.

Environment Canada. *Preliminary Proceedings: Municipal Waste Incineration.* October 1-2, 1987.

Environment Canada. "The National Incinerator Testing and Evaluation Program: Air Pollution Control Technology (Quebec City)." September, 1986.

Environmental Defense Fund (Richard A. Denison and John Ruston, editors). *Recycling & Incineration: Evaluating the Choices.* Washington, DC: Island Press, 1990.

Environmental Defense Fund. *The Hazards of Ash and Fundamental Objectives of Ash Management.* New York: 1989.

Environmental Defense Fund. *To Burn or Not to Burn: The Economic Advantages of Recycling Over Garbage Incineration for New York City.* New York: August, 1985.

Franklin Associates, Ltd. "Characterizing the Waste Stream." Prepared for the US Environmental Protection Agency. May 20, 1988.

Gershman, Brickner, and Bratton, Inc. *Small-Scale Municipal Solid Waste Energy Recovery Systems.* Van Nostrand Reinhold Company, Inc., 1986.

Hahn, Jeff, and Donna Sofaer. "Variability of NO_x Emissions from Modern Mass Fired Resource Recovery Facilities." Prepared for the 81st Annual Meeting of Air Pollution Control Association, June 19-24, 1988.

Hang, Walter Liong-Ting and Steven A Romalewski. *The Burning Question: Garbage Incineration Versus Total Recycling in New York City.* The New York Public Interest Research Center: 1986.

Herstad, Solvie, and A. Kullendorf. "Waste Incineration by Fluidized Bed Technology — Test Results and Experience." Proceedings Municipal Waste Incineration, Environment Canada, National Ininerator Testing and Evaluation Program. Montreal: October 1-2, 1987.

Industrial Gas Cleaning Association. Conference Proceedings of IGCI Forum '88 and IGIC Forum '90, Washington, DC.

INFORM (Maarten de Kadt). "Recycling Programs in Somerset County, New Jersey, and Islip, New York." New York: in press.

INFORM (Maarten de Kadt). "Managing Westchester's Garbage: Building on Experience." *Westchester Environment.* Summer, 1990. Federated Conservationists of Westchester County.

INFORM (Marjorie J. Clarke). *Technologies for Minimizing Emission of NO_x from MSW Incinerators.* New York: 1989.

INFORM (Marjorie J. Clarke). *Improving Environmental Performance of MSW Incinerators.* New York: November, 1988.

INFORM (Allen Hershkowitz and Eugene Salerni). *Garbage Management in Japan: Leading the Way.* New York: 1987.

INFORM (Allen Hershkowitz). *Garbage Burning: Lessons from Europe: Consensus and Controversy in Four European States.* New York: 1986.

Institute for Local Self-Reliance (Brenda Platt, *et al.*). *Garbage in Europe: Technologies, Economics, and Trends.* May 1988.

Interpoll Laboratories Report. "Results of the November 3-6, 1987 Performance Test on the No. 2 RDF and Sludge Incinerator at the WLSSD plant in Duluth, Minnesota."

Kocher, Peg and Anita Siegenthaler. *The World of Waste.* League of Women Voters of the Tri-State Metropolitan Region: New York, 1988.

Lauber, Jack D., and Donald A. Drum. "Best Control Technologies for Regional Biomedical Waste Incineration." Prepared for the 83rd Annual Meeting of the Air and Waste Management Association, June 27, 1990.

Linak, W. P., *et al.* "Waste Characterization and the Generation of Transient Puffs in a Rotary Kiln Incinerator Simulator." Prepared for the 13th Annual Research Symposium on Land Disposal, Remedial Action, Incineration, and Treatment of Hazardous Waste. Cincinnati, Ohio: July, 1987.

Mannis, Barry A. *Waste to Energy: Cash From Trash.* Shearson Lehman Hutton. 1986.

McDaniel, M. D., *et al.* "Air Emissions Tests at Commerce Refuse-to-Energy Facility — May 26-June 5, 1987." Energy Systems Associates for County Sanitation Districts of Los Angeles County.

MRI Project. "Results of the Combustion and Emissions Research Project at the Vicon Incinerator Facility in Pittsfield, Massachusetts." June 3, 1987.

National Solid Wastes Management Association. "Landfill Capacity in the Year 2000." 1989.

National Solid Wastes Management Association. *Landfill Capacity in the US: How Much Do We Really Have?* October, 1988.

Natural Resources Defense Council, Environmental Defense Fund, INFORM, Environmental Action Coalition, Scenic Hudson. *A Solid Waste Blueprint for New York State.* New York: March, 1988.

Needleman, Herbert, *et al.* "Deficits in Psychological Classroom Performances of Children with Elevated Dentine Lead Levels," *New England Journal of Medicine,* vol. 300, 1979.

Sommer, Edward J., *et al.* "Emissions, Heavy Metals, Boiler Efficiency, and Disposal Capacity for Mass Burn Incineration with a Presorted MSW Fuel." Prepared for the 81st Annual Meeting Air Pollution Control Association, June 19-24, 1988.

Taylor, Hunter F. *Energy Recovery from Municipal Solid Waste.* Energy Division, Office of Emergency and Energy Services. Commonwealth of Virginia: June, 1984.

Toxic Substance Control Commission. "Recommendations for Policy and Regulations for Residue from MSW Incineration." Michigan: August, 1988.

US Conference of Mayors and National Resource Recovery Association. *City Currents.*

US Environmental Protection Agency, Office of Solid Waste. "Characterization of Municipal Solid Waste in the United States: 1990 Update. June, 1990.

US Environmental Protection Agency. Conference Proceedings of International Conference on Municipal Solid Waste Combustion. Hollywood, Florida: April 11-14, 1989.

US Environmental Protection Agency. *The Solid Waste Dilemma: An Agenda for Action.* February, 1989.

Waste Age. November, 1990.

Waste Not. Vol. 37, No. 40. January, 1989.

Zimmermann, Elliott. *Solid Waste Management Alternatives: Review of Policy Options to Encourage Waste Reduction.* Illinois Department of Energy and Natural Resources. February, 1988.

Appendix D. Glossary

Acid gases. A group of gases with acidic properties, including sulfur dioxide, hydrogen chloride, hydrogen fluoride, and oxides of nitrogen, that form during combustion from sulfur, chlorine, fluorine, and nitrogen in garbage.

Activated carbon. Finely ground carbon particles treated to permit adsorption of pollutants in internal pore spaces.

Arch. A constrained entrance to the furnace which, if carefully placed, can slow the flow of air from the grate into the furnace proper, thus enhancing combustion efficiency. See also **bullnoses.**

Attainment area. A region within the United States is said to be in attainment if it meets the ambient air concentration standards established by the federal Environmental Protection Agency for one or more of the six criteria pollutants. An area may be in attainment for one criteria pollutant and not for others. See also **nonattainment area.**

Auxiliary burner. A burner located in the furnace that burns a fuel other than municipal solid waste (such as natural gas or oil) during startup, shutdown, and temperature upsets in an incinerator, thereby stabilizing combustion (and minimizing creation of products of incomplete combustion) by maintaining a minimum furnace temperature.

Averaging time. The amount of time over which emissions are averaged.

BACT. See **best available control technology.**

Baghouses (also called **fabric filters**). A state-of-the-art particulate-removal technology consisting of large structures containing woven fabric bags that work much like vacuum cleaner filters, passing the air through while capturing the particulates.

Batch stoking. A system for introducing waste into an incinerator, in discrete batches, usually by a front-end loader truck in combination with a ram or pneumatic feeding device. See also **continuous stoking.**

Best available control technology (BACT). A policy for achieving the maximum degree of emission reduction of regulated pollutants in a flexible way that includes energy, environmental, and economic impacts, as well as other cost considerations, allowing regulators to continually redefine what is an attainable and enforceable emissions standard. See also **lowest achievable emissions rate.**

Boiler ash. Incompletely burned material that leaves the furnace suspended in combustion gases and falls out of the gases in the boiler as the gases cool. See also **fly ash** and **bottom ash.**

Bottom ash. Non-airborne unburned solid matter that falls through the grate and accumulates at the bottom of an incinerator. See also **fly ash** and **boiler ash.**

British thermal units (BTUs). The unit of heat energy required to raise the temperature of 1 pound of water by 1°F.

BTU values. A measure of the amount of heat energy generated when a given amount of materials are burned, using British Thermal Units.

Bullnoses. Protrusions from the furnace wall which, if carefully placed, cause turbulence in the flow of air from the grate, thus enhancing combustion efficiency. See also **arch.**

Carbon dioxide (CO_2). A colorless, odorless, nontoxic gas that results from combustion. Although carbon dioxide is a normal component of the ambient air, increased CO_2 emissions are thought to contribute to global climate change.

Carbon monoxide (CO). A colorless, odorless gas that interferes with the blood's ability to absorb oxygen; it is a product of incomplete combustion.

Cementation. An incinerator ash treatment process that involves mixing the ash with cement to create a hard mass with less leaching potential. See also **fixation.**

Char. Unburned carbon-containing materials that remain on the grate following the primary combustion phase.

Combustion. Burning; the process by which wastes are broken down, in the presence of heat and oxygen, thereby releasing energy in the form of heat and light. Ideally, combustion produces only carbon dioxide and water vapor, but, in reality, many more elements and compounds result.

Composting. A natural aerobic process involving the biological decomposition of organic wastes by microorganisms, producing a humus-like material (compost) that can be used as a landfill cover and, if the metal content is low, in agriculture.

Continuous emissions monitors (CEMs). Monitors that track emissions of gases such as carbon dioxide, carbon monoxide, hydrogen chloride, sulfur dioxide, and oxides of nitrogen, as well as opacity, on an ongoing basis so that corrective measures can be implemented, if needed, in a timely fashion.

Continuous process monitors (CPMs). Monitors that track such incinerator processes as furnace temperature, oxygen content, flue gas temperature, steam pressure, and steam flow on an ongoing basis so that corrective measures can be implemented, if needed, in a timely fashion.

Continuous stoking. A system for introducing waste into an incinerator a few tons at a time (usually down an inclined chute from a crane) so that waste is fed without interruption into the furnace. See also **batch stoking**.

Criteria pollutants. Pollutants for which the federal Environmental Protection Agency, as mandated by the 1970 Clean Air Act, has established maximum allowable air concentration limits based on an evaluation of their potential health and environmental efforts. There are six: ozone, sulfur dioxide, oxides of nitrogen, carbon monoxide, lead, and particulates.

Cyclone. A particulate control device that funnels flue gases into a spiral, creating a centrifugal force that removes larger particles.

Design capacity. The maximum amount of fuel an incinerator is designed to burn.

Dioxins. A class of 75 polychlorinated organic compounds with very similar chemical properties, some of which are known to be highly toxic to animals. See also **furans**.

Dry injection scrubbers. A kind of scrubber that injects, into the flue gas, dry powdered lime or another alkaline agent that reacts with acid gases.

Dual-chambered furnace. A furnace containing a primary chamber in which primary combustion takes place and a separate chamber for secondary combustion; they usually burn a smaller quantity of garbage than single-chambered furnaces and are less common. See also **single-chambered furnace**.

Eadon toxic equivalents. Units that give a comparative indication of toxicity levels of dioxin and furan emissions; they are obtained by converting measured emissions of different dioxins and furans into a standard format that takes into account differing toxicity levels of different dioxins and furans.

Economizer. The last heat-removing section of a boiler.

Electrostatic precipitators (ESPs). Particulate control devices consisting of one or more pairs of electrically charged plates or fields; particulates in the flue gas are given an electrical charge, forcing them to be drawn out of the gas stream to stick to the plates. Electrostatic precipitators with four or more fields (or with two or three fields with especially large collection areas) are considered state-of-the-art equipment.

Emissions. Products that form in an incinerator and are discharged from the stack into the air as gases or small particles.

Energy rating. The maximum amount of energy a waste-to-energy plant can produce.

Entrained. Collected and transported by the flow of air moving at high velocity; used to describe particulates lifted off the grate and carried upwards into combustion gases during combustion.

ESPs. See **electrostatic precipitators**.

Excess air. Air in the furnace system that is not actually required for combustion.

Extraction Procedure Toxicity Test (EP Tox). A laboratory test designed to determine the likelihood that certain metals and other constituents could be leached from incinerator ash in a landfill by acid.

Fabric filters (also called **baghouses**). A state-of-the-art particulate-removal technology consisting of large structures containing woven fabric bags that work much like vacuum cleaner filters.

Ferrous metals. Iron-containing metals (including steel).

Fixation. An incinerator ash treatment process that involves mixing the ash with cement and/or alkaline scrubber materials to create a hard mass with less leaching potential. See also **cementation.**

Flow-control ordinance. A regulation that guarantees a constant supply of waste to a disposal facility. Such ordinances declare garbage the property of the municipality once it is placed at curbside, enabling the municipality to ensure that the garbage will be used to feed the incinerator. These ordinances are also used in municipal recycling programs.

Fluidized bed combustor. A promising new furnace design in which processed waste is injected into a loose bed of sand and limestone particles that are in a constant state of turbulence; air passing through the bed reacts with the heated refuse-derived fuel and its combustion products.

Fly ash. Incompletely burned materials in solid or condensible form that leave the furnace suspended in combustion gases and are subsequently trapped in emission control devices. See also **bottom ash** and **boiler ash.**

Furans. A class of some 135 polychlorinated organic compounds with similar chemical properties, some of which are known to be highly toxic to animals. See also **dioxins.**

Heavy metals. Metals with high atomic weights (such as lead, cadmium, chromium, mercury, and arsenic); many can be toxic at low concentrations and can accumulate in the food chain.

Hydrogen chloride. An acid gas that forms during garbage incineration when chlorine present in waste combines with hydrogen from water vapor.

Hydrogen fluoride. An acid gas that forms during garbage incineration when fluorine present in waste combines with hydrogen from water vapor.

LAER. See **lowest achievable emissions rate.**

Leachate. Liquid containing dissolved substances formed by water trickling through wastes, agricultural pesticides or fertilizers, or other materials.

Liner. A barrier, typically made from clay or plastic, designed to prevent leachate from leaking from a landfill.

Lowest achievable emissions rate (LAER). A policy for limiting emissions of regulated pollutants that requires plants to use the best demonstrated technology based solely on environmental considerations, regardless of cost. See also **best available control technology.**

Mass burn incinerators. Incinerators that burn garbage as received with little attempt on-site to separate objects that may not burn well or burn at all. See also **refuse-derived fuel incinerators.**

Materials recovery facility (MRF). A waste-processing plant that processes source-separated or mixed recyclables (such as paper, plastics and glass) into individual materials available for market. The separated materials are kept whole and sold.

Micron. A unit of length measurement: 1 micron equals 1 micrometer (10^{-6} meters) equals 1/25,000 of an inch. Used to measure particle sizes.

Monofill. A landfill containing only one material, usually used to refer to ash-only landfills.

MRF. See **materials recovery facility.**

Multiple-chambered furnace. A furnace containing four chambers, used for medical waste incinerators, but not yet for garbage incinerators. The third and fourth chambers allow for additional combustion. See also **single-** and **dual-chambered furnace.**

Municipal solid waste. Garbage collected from the residential, commercial, and institutional sectors.

Municipal waste combustors (also called **waste-to-energy plants** or **resource recovery plants**). Incinerators that recover heat energy from burning garbage; the energy, in the form of steam, can be circulated for heating or converted to electricity.

Nonattainment area. A region within the United States is said to be in nonattainment if it does not meet the air concentration standards established by the federal Environmental Protection Agency for one or more of the six criteria pollutants. See also **attainment area.**

New Source Performance Standards (NSPS). The first comprehensive national incinerator regulations, issued by the federal Environmental Protection Agency in 1991. Also known as Standards of Performance for New Stationery Sources (Municipal Waste Combustors).

Opacity. The amount of light obscured by particulates in the air. It is used as an indicator for determining the level of particulate emissions from an incinerator.

Overfire air (also called **secondary air**). Air injected above the grate during the secondary phase of combustion.

Oxides of nitrogen. A group of acid gases (including nitrogen oxide [NO] and nitrogen dioxide [NO_2]), collectively termed NO_x, that form during garbage incineration when nitrogen from the wastes and/or from the atmosphere combines with oxygen from the air. Key contributors to ozone/smog and acid rain. Criteria pollutant.

Particulates. Minute particles in solid or liquid form produced during incineration of municipal or solid waste. Particulates range in size from more than 500 microns to less than 0.1 micron in diameter. Criteria pollutant.

Pollutant precursors. Elements or compounds present in solid waste which, when burned, are transformed into emissions.

Primary air (also called **underfire air**). Air injected into the furnace, generally from below the fire, during the primary phase of combustion when garbage is first exposed to the flames.

Primary combustion phase. The first phase of incineration, during which burning garbage is transformed into bottom ash or char, with volatile gases and incompletely burned carbon compounds also produced. See also **secondary combustion phase.**

Products of incomplete combustion. A variety of carbon compounds, including carbon monoxide and dioxins and furans, that are produced when garbage does not completely burn.

RDF. See **refuse-derived fuel.**

Recycling. A process by which materials (such as paper, glass, metal, and plastics) that would otherwise be disposed of as waste are separated, collected, processed, and remanufactured into new products.

Refractory walls. Ceramic, brick, and stone walls surrounding a furnace that reflect heat back into the fire, thus keeping the furnace exterior cool. See also **water walls.**

Refuse-derived fuel (RDF). The combustible material left after municipal solid waste is sorted (with recyclable and noncombustible materials removed); sometimes processed to a small uniform size (pellets), and sometimes left as fluff.

Refuse-derived fuel incinerators. Dedicated incinerators that burn wastes that have been processed and sorted, with recyclable and noncombustible materials removed. See also **refuse-derived fuel** and **mass burn incinerators.**

Residence time. The time during which combustion gases are retained in the furnace during the secondary combustion phase; one of the **three T's** involved in maximizing combustion efficiency.

Resource recovery plants (also called **municipal waste combustors** or **waste-to-energy plants**). Incinerators that recover heat energy from burning garbage; the energy, in the form of steam, can be circulated for heating or converted to electricity.

Retrofitting. The process of redesigning an existing facility to meet new standards and/or use new technologies.

Rotary combustor. A furnace design consisting of a large, slightly inclined, rotating cylindrical furnace. The furnace is designed to improve combustion efficiency by allowing for continuous mixing of waste on the grate and exposure of new surface areas to heat and air.

Scrubbers. A group of devices, some of which are the state-of-the-art equipment for controlling emissions of acid gases, using techniques such as condensation and acid/base reactions to neutralize acid gases in flue gases. Scrubbers also play a role in reducing emissions of dioxins and furans, oxides of nitrogen, and mercury. See also **wet scrubbers, spray-dry scrubbers**, and **dry-injection scrubbers**.

Secondary air (also called **overfire air**). Air injected above the grate during the secondary phase of combustion. See also **overfire air.**

Secondary combustion phase. The phase of combustion during which the gases formed in the primary phase rise above the grate and are themselves oxidized. See also **primary combustion phase**.

Selective catalytic reduction. A system for neutralization and removal of oxides of nitrogen from flue gases that operates by injecting ammonia into flue gas after it passes through particulate control devices and before a catalyst bed consisting of metallic materials in a variety of forms. See also **selective noncatalytic reduction**.

Selective noncatalytic reduction. A system for neutralization and removal of oxides of nitrogen from flue gases. Among several variations of this technology, one uses injection of ammonia into the furnace and one injection of aqueous urea into the furnace and boiler. See also **selective catalytic reduction**.

Single-chambered furnace. The most common type of incinerator, usually designed for processing 250 or more tons of waste per day; both phases of combustion take place in the same furnace. See also **dual-** and **mulitple-chambered furnace**.

Source reduction. Reducing the amount and toxicity of garbage generated in the first place.

Source separation. Separation of different materials (such as papers, metals, plastics, glass) in municipal solid waste at the source; that is, in the home, or at curbside, before the garbage is picked up for disposal.

Spray-dry (or **semi-dry**) **scrubbers.** A kind of scrubber that captures acid gases by impaction of the gas molecules onto an alkaline slurry such as lime.

State of the art. As used in this book, the best current technologies and techniques for regularly achieving reductions in the environmental impacts of waste-to-energy plants.

Sulfur dioxide. A pungent, colorless acid gas that forms during combustion when sulfur present in garbage combines with oxygen from the air. Key contributor to acid rain; criteria pollutant.

Telemetering. Instantaneous computer transmission of continuous monitoring data to local or state authorities.

Three T's. The three factors — temperature, turbulence, and time — involved in maximizing combustion efficiency during the secondary combustion phase.

Tipping fee. The fee charged to dispose of solid waste at an incinerator, landfill, or other waste processing facility.

Tipping floor. The surface onto which waste entering a garbage-burning plant is dumped.

Turbulence. Adequate mixing of the combustion gases with oxygen.

Underfire air (also called **primary air**). Air injected into the furnace, underneath the grate, during the primary phase of combustion when garbage is first exposed to the flames.

Vitrification. An incinerator ash treatment process that involves quickly cooling heated ash to form an impermeable, glassy product.

Waste-to-energy plants (also called **municipal waste combustors** or **resource recovery plants**). Incinerators that recover energy from burning garbage; the energy, in the form of steam, can be circulated for heating or converted to electricity.

Water walls. Furnace walls with pipes containing constantly circulating water that absorbs heat from the furnace, and transmits it to the heat-recovery boiler, thus keeping the exterior furnace walls cool. See also **refractory walls.**

Wet scrubbers. A kind of scrubber that captures acid gases by condensation of the gas molecules onto water droplets, sometimes with alkaline agents added in small amounts to aid in the reaction.

INDEX

Note: An *italic page number* indicates a definition of the term. The letter "*t*" after a page number indicates that the information is presented in a table; the letter "*n*" indicates a footnote. The names in all capital letters refer to plants in the study.

A

B

C

G

H

I

J

L

R

S

T

U

V

W

Y

About the Authors

Marjorie J. Clarke

Marjorie Clarke joined INFORM in April, 1988, and now serves as a consultant to INFORM's Municipal Solid Waste Program. She currently chairs the Air and Waste Management Association's Technical Committee on Municipal Solid Waste and the Waste Planning Committee of the Manhattan Citizen's Solid Waste Advisory Board.

Ms. Clarke's environmental career began with an internship at the United States Environmental Protection Agency in 1974. Her solid waste management experience since then has included a variety of positions, including policy coordinator for resource recovery for the New York Power Authority and environmental scientist for the New York City Department of Sanitation's Office of Resource Recovery.

She holds two masters degrees: an M.S. in applied sciences from New York University, and an M.A. in geography and environmental engineering from Johns Hopkins University.

Maarten de Kadt, Ph.D.

Maarten de Kadt joined INFORM in March, 1987, as a Research Associate in the Municipal Solid Waste Program.

Dr. de Kadt was a key contributor to INFORM's *Business Recycling Manual,* copublished with Recourse Systems, Inc. He also prepared INFORM's forthcoming analysis of recycling programs in the Town of Islip (New York) and Somerset County (New Jersey) and wrote an article, "Managing Westchester's Garbage: Building on Experience," that appeared in the *Westchester Environment.* Dr. de Kadt was a coauthor of two INFORM studies addressing the solid waste crisis: *Garbage: Practices, Problems, & Remedies* and *Solid Waste Management: The Garbage Challenge for New York City.*

Prior to joining INFORM, Dr. de Kadt taught at Lehman College, Wagner College, and Empire State College. He earned his Ph.D. in economics from the New School for Social Research and his M.B.A. in marketing research from Baruch College.

David Saphire

David Saphire joined INFORM's Municipal Solid Waste Program as a Researcher in January, 1989.

Mr. Saphire has conducted research on Dutchess County's (New York) solid waste planning and has written and presented testimony on that county's proposed recycling law. He has developed an analytical framework for assesing recycling programs, and is currently researching municipal and corporate source reduction initiatives in the United States.

Previously, Mr. Saphire worked as a tenant-landlord mediator for the New York State Division of Housing and Community Renewal and as an environmental technician at the United States Testing Company.

He received his B.A. in environmental science from the State University of New York at Binghamton.

About the Editor

Sibyl R. Golden
Sibyl Golden joined INFORM in September, 1989, and is now Director of Research and Publications. She is coauthor of INFORM's Special Report, *Toxic Clusters: Patterns of Pollution in the Midwest.*

Prior to joining INFORM, Ms. Golden was a science editor and writer, working at McGraw-Hill and several journal publishing companies. She also held a variety of communications, community relations, and management positions at the Port Authority of New York and New Jersey.

She earned her A.B. *cum laude* from Harvard University, where she majored in biology.

INFORM Publications

Selected Publications on Municipal Solid Waste

Business Recycling Manual
(copublished with Recourse Systems, Inc.). 1991, 202 pp., $85.00.

Garbage Management in Japan: Leading the Way
(Allen Hershkowitz, Ph.D., and Eugene Salerni, Ph.D.). 1987, 152 pp., $15.00.

Garbage Burning: Lessons from Europe: Consensus and Controversy in Four European States
(Allen Hershkowitz, Ph.D.). 1986, 64 pp., $9.95.

Technologies for Minimizing the Emission of NO_x *from MSW Incinerators*
(Marjorie J. Clarke). 1989, 33 pp., $9.95

Improving Environmental Performance of MSW Incinerators
(Marjorie J. Clarke). 1988, 82 pp., $15.00.

Solid Waste Management: The Garbage Challenge for New York City
(Maarten de Kadt, Ph.D., and Nancy Lilienthal). 1989, 56 pp., $7.95.

Forthcoming Publications on Municipal Solid Waste

Reducing Office Paper Waste (working title, in preparation)

Planning for Source Reduction (working title, in preparation)

Other INFORM Publications

INFORM also publishes reports on chemical hazards prevention, urban air quality, and land and water conservation, and a quarterly newsletter. For a complete publications list and more information, call or write to INFORM (address on next page).

SALES INFORMATION

Payment

Payment, including shipping and handling charges, must be in US funds drawn on a US bank and must accompany all orders. Please make checks payable to INFORM and mail to:

INFORM
381 Park Avenue South
New York, NY 10016

Shipping Fees

To order in the US, please send a check that includes $3.00 for the first book and $1.00 for each additional book for shipping and handling charges. To order in Canada, add $5.00 for the first book and $3.00 for each additonal book. For information on shipping rates for other countries, call (212) 689-4040.

Discount Policy

Booksellers:	20% on 1-4 copies of same title	
	30% on 5 or more copies of same title	
General bulk:	20% on 5 or more copies of same title	
Public interest and community groups:	Books under $10:	no discount
	Books $10-$25:	$10
	Books $25 and up:	$15

Returns

Booksellers may return books, if in saleable condition, for full credit or cash refund up to 6 months for date of invoice. Books must be returned prepaid and include a copy of the invoice or packing list showing invoice number, date, list price, and original discount.

Board of Directors

	DATE DUE		
JUN 0 2 1993			